My Lively Lady

My
Lively Lady

by ALEC ROSE

NAUTICAL PUBLISHING COMPANY

K. ADLARD COLES   COMMANDER ERROLL BRUCE RN (RTD)

Captain's Row · Lymington · Hants

in association with

GEORGE G. HARRAP & COMPANY LTD.

London · Toronto · Wellington · Sydney

Standard edition SBN 245 59565 1
Limited edition SBN 245 59593 7

First published in Great Britain by
NAUTICAL PUBLISHING COMPANY
Captain's Row, Lymington, Hampshire

*Composed in 11 on 12 pt Monotype Baskerville
and made and printed in Great Britain by
Cox & Wyman Ltd., London, Reading and Fakenham*

Acknowledgements

My thanks are firstly due to my wife Dorothy, who made my voyages possible. For her it was not merely a matter of keeping the business going while I was away; it was her encouragement which made the whole thing possible. I also wish to thank the committee of our friends who looked after my affairs after my voyage came into the news at Melbourne: Noel (Rear Admiral) Clarke, Roland Phillips, Douglas Phillips-Birt and (Sir) David Mackworth who advised me on rigging and flew out to help me when *Lively Lady* arrived at Bluff, with a damaged masthead fitting. Nor must I forget my professional advisers, Peter Carter-Ruck and Phillip Woolley.

For those who helped me in the voyage itself I am specially grateful. I have mentioned some of them in this book, particularly the Mashford family at Plymouth, the Royal Yacht Club of Victoria and the Mayor of Williamstown who made our stay so enjoyable, and Mr. Chisholm Cutts who replaced my broken rigging; the Bluff Engineering Company, the Mayor of Bluff and the Manager, New Zealand Stevedoring Company, who helped me so well; the Commander-in-Chief and the Queen's Harbour Master who welcomed me on arrival at Portsmouth. There were so many helpers and supporters in England, Australia and New Zealand that I cannot acknowledge all the kindnesses I received and all the friends I made. All I can say here is that I shall always treasure the warmth of heart which was everywhere shown to me.

Turning to this book, I wish first of all to thank those who have written the chapters in Part II: Radio Officer Roberts, Michael Mason, Erroll Bruce, and, of course, Dorothy for telling what went on while I was away; for the technical appendices I thank Douglas Phillips-Birt and 'Blondie' Hasler. The sectional drawings of *Lively Lady* were contributed by Charles Hurford. The lines of *Lively Lady* were kindly supplied by the designer, Frederick Shepherd and the tracing by the *Yachting Monthly*; the diagrams of the self-steering gear are by the courtesy of M. G. Gibb Ltd., the maps and end paper by Keith Blount. My introduction to *Why Alone* is based on an article which appeared in the *Yachting World Annual* and Chapter 3 on the Transatlantic Race was first published in *The Yachtsman*; both appear here by kind permission of the respective Editors. I also thank the ladies of Lymington who typed the manuscript. I have never met them personally, but I appreciate the way in which they produced the typewritten sheets within a few hours of receiving my handwritten manuscript in pencil. I believe one of them also provided me with a pencil sharpener. Neither must I forget my publishers Adlard Coles and Erroll Bruce, especially Adlard Coles and his wife for kindly putting their quiet home at my disposal where I could write in peace.

For the photographs, I have pleasure in acknowledging the following: *The Age* Melbourne, plates 7, 8, 14, 16; Central Office of Information, plate 34; Del Mar Studios, Invercargill, plates 18, 19 and 20; *Herald-Sun*, Melbourne, plate 15; I.P.C. Syndicate 33; David Lolley, of *Portsmouth Evening News*, 1, 9, 10, 11, 29, 30, 31, 32, 35; Ministry of Defence, 22, 25, 28; Press Association, 26; J. Roberts, 24; *Sport and General*, 27; Thomson Newspapers Limited (Topix) 17. I regret my own photographs were failures except Plate No. 23, but these are more than compensated for by the magnificent pictures contributed by Captain Th. de Lange which appear on plates 13, 21 and the front end paper.

Contents

Continued on page x

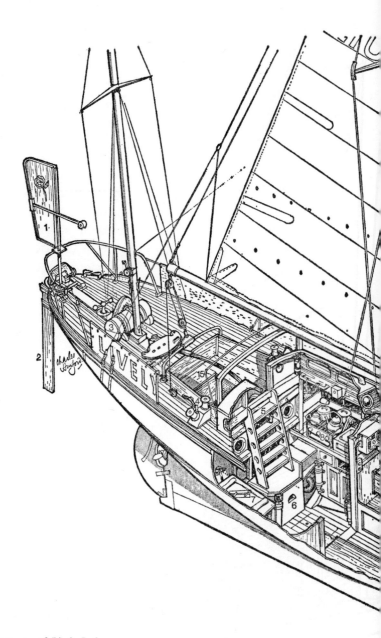

Cutaway of *Lively Lady*

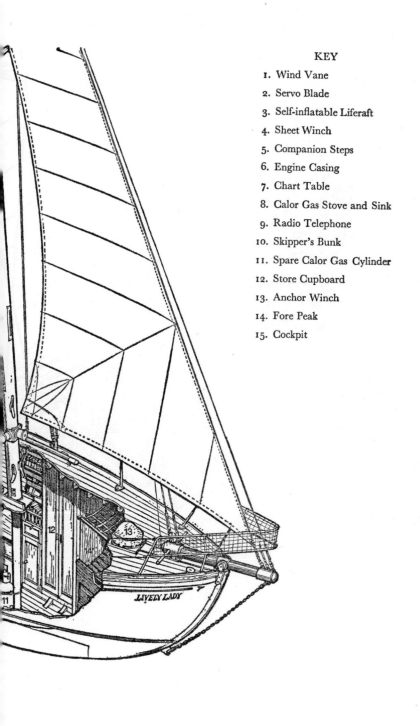

KEY

1. Wind Vane
2. Servo Blade
3. Self-inflatable Liferaft
4. Sheet Winch
5. Companion Steps
6. Engine Casing
7. Chart Table
8. Calor Gas Stove and Sink
9. Radio Telephone
10. Skipper's Bunk
11. Spare Calor Gas Cylinder
12. Store Cupboard
13. Anchor Winch
14. Fore Peak
15. Cockpit

PART II

APPENDICES

List of Plates

Maps and Drawings

Introduction - Why Alone?

W<small>HEN</small> I first entered the single-handed Transatlantic Race and made the return passage on my own I remember a comment that was often made:

'To cross the broad stormy Atlantic alone – you must be mad.'

Yes indeed, 'Why alone?'

How often have I been asked that question, together with, 'Weren't you lonely?' 'Were you ever frightened?' 'What did you do with yourself all the time?' And so on. The answer to most of these queries is easy and logical; but to answer the question 'Why alone?' is harder. At least I know why – but it is difficult to put into words.

To start at the beginning. I suppose one must be basically of a certain type; a sort of 'dark horse', if you like; a thinker; a dreamer; an idealist; an individualist. A man prepared to stand or fall by his own decisions. All these descriptions would fit in some degree the single-handed sailor – as it would the lone explorer or the lone trapper of the frozen North. Let me say at once that they are not odd people who are anti-social. Indeed, they are usually jolly good company at a party. But they have a sense of values in life and although money is essential to achieve their ambitions, it is not money in itself that is important. They do not expect to make money.

In my own case, from an early age I was considered to be not very strong and was unable to go in for sports with my school-fellows as I wished. By my 'teens I had recovered, though I was dreadfully shy and often afraid to do things in front of my elders

that in my mind I knew I could do – which made me rather a
lone soul. I loved to go on long country walks well off the beaten
track and creep along a hedgerow to watch Nature at work. To
lie quite still on a sunny bank and listen to the fantastic
noises going on around is wonderful. Apart from the variety of
birds not only singing, but talking to each other, there are
thousands of winged insects humming from plant to plant, or
the crawling varieties each busily going about the business of
living and surviving. All this, added to the rustle of the leaves
or the gurgle of a stream, creates real din. How can one feel
lonely in such company? It is there for us to look at, to listen to
and to enjoy. We are all part of it. This applies equally to the
sea, as I will try to explain later. I dreamed of far-off places I
had read about. I would lie on the cliff-top and watch the smoke
of the distant steamers passing down channel and wonder where
they were bound.

It is small wonder then that I have always been one to venture
out on my own. I am not at my best in a crowd. This applied
equally in my business life and I scorned a safe job and regular
hours for the long and irregular hours of being self-employed.
This has its compensations when one wants to undertake a
venture such as my first lone trip across the Atlantic. For years
I had saved and prepared myself for it, sacrificing other plea-
sures or luxuries and keeping my ambitions to myself. Not that I
am secretive by nature. I just felt it was my personal affair and
really interested no-one else. For years – literally years – I prepared the yacht for the job. I served during the War on the
North Atlantic convoys, and having a very vivid imagination I
could visualize the worst that could happen to the yacht, such
as being rolled over by huge seas, and I had the fittings,
rigging, and gear to withstand it – that is, material two or three
times as strong as normally required in a cruising yacht such as
mine. You see, I wanted to survive the ultimate, the knock-
down; when the sea had taken its toll of the less prepared I
hoped to be one of those still left in a seaworthy condition. All this
I had gone over and over in my dreams – before starting to fit out
my yacht – and so when the time came I was mentally ready.

When the first single-handed Transatlantic Race was organ-
ized in 1960 I was unprepared. The second race came along in

1. The author on board his *Lively Lady*.

2. The author's eight horse ploughing team when working in Canadian prairies and below:

3. Mounted on one of the team to round-up cattle.

The author holds his eldest son, ec, at their home in Canterbury, 1932.

5. The author's brother, Dennis, seen when both were working in their father's haulage business in Canterbury.

6. The author with his eldest son, Alec, and his three grand-children, before the start of the Transatlantic race 1964.

7. (above) Porridge with a dash of whisky for breakfast and 8. (below) checking the sextant which finds the position at sea.

9. *Lively Lady* sails round from Langstone Harbour to Southsea before starting round the world.

10. The author's youngest daughter Jane waves farewell at Spithead.

11. Alec and Dorothy Rose, with Algy, on board *Lively Lady*.

12. Vice-Commodore Graeme West and Commodore John Hayward of the Royal Yacht Club of Victoria with the author and the Governor of Victoria, Sir Rohan Delacombe.

1964 and I threw everything into it. Apart from anything else I realized I was getting older. It was a strain financially, but if I had waited until my accountant said I could afford it I knew I should never get away.

I became more and more excited as the great day drew nearer and my dreams became fact. My friends would come along and ask if it were true and express real concern. Quite a few believed they would never see me again. They all thought I was completely mad. For myself I was excited, but in a quietly confident way. I was also so proud, and yet very humble, to be numbered among those whom I admired so much, men like Chichester, Howells, and Hasler. How could one expect to beat them? I didn't. But I did expect to finish the course. In fact I hardly gave a second thought to that. The months before the race was the time to think and worry – going over all the details again and again.

The cocktail parties and dinners we attended in Plymouth were designed, I think, to test our stamina before the start. I for one was not feeling too bright as we left our moorings on the morning of the 23rd of May, and was relieved that it was not blowing too hard as the gun went. The race was now on, though no-one might have known it on Sunday afternoon as some of us lay becalmed off Land's End. By Monday morning, however, I was clear of the Scillies and heading west in a fresh breeze. I now felt fine. But 'Why alone?' you ask again.

My friends, I was happy and content. I had all I needed. I had a good ship under me and I felt as free as the birds that circled above. I was king of my little world. I walked round the decks and admired everything. Then I looked at the sea. It was boisterous and playful. I admired it, but respected it, acknowledging 'King Neptune' as king of the sea. Would he allow me to remain king of my sturdy yacht? Time would tell; but he would test me for sure, so I went about my duties in a slightly more sober mood. The sea is a great leveller and quickly humbles the big-headed sailor. I became acutely aware of being on my own, alone making all decisions, choosing the course to steer, deciding what sails to carry, fixing the position by sun-sights, writing up the log, preparing food, and whenever possible washing and shaving, resting and sleeping. I think that answers the question 'What did you do with yourself all the time?' One day I wrote in my log: 'Feel tired. Haven't sat down all day – rather as we hear the housewife say in the evening.' Was I lonely? To that I will answer 'alone' but never 'lonely'. One can be lonely in a big city – though not alone. The sea is alive in its different moods.

On the Atlantic Race there were always birds of some sort and I watched their effortless glide with hardly a wing movement. Then there were the small stormy petrels with their sharp dives, after the style of a swallow, as they picked out their food from the frothy foam of a breaking crest when it creamed down the face of the wave. One of the happiest moments too was the dawn. After a cold dark night I would doff my hat and bid the sun 'Good morning!' One realizes what a warm life-giver he is, what a great morale-raiser. And there was nothing to be lonely about when I heard a great deep-throated sound and I watched the graceful movements of a whale right alongside. But I was fearful of his great strength.

Not 'lonely' either when a great company of dolphins splashed me as they leaped and dived all around me keeping station on the yacht as we sailed along. They were often with me and were great company.

Was I frightened? Yes, often, though perhaps worried would be the better word. As time passed and I weathered gales that were at times severe I became more and more confident that the yacht was capable of riding them so long as I didn't fail her. This was what I was concerned about mostly – that I was doing the yacht justice. To leave the comparative safety of the cockpit and crawl forward to a heaving and waveswept foredeck is at times rather frightening. But on completion of the task and safely back again, one gets a terrific boost at having proved oneself capable of wrestling with the elements. I would then acknowledge the might and power of the sea and ask him to calm down and let us be friends. Again, when on the homeward journey and approaching the Scilly Isles I rode out a Force 10 gale. I wrote in my log at the time: 'What a terrible black stormy night it is – God help us – I suppose I shouldn't worry but I'm a little anxious.' I suppose one could call that being a little frightened. But I was always able to tackle the next job that came along and nothing got out of hand.

My most fearful moments, though, were when we were sailing along in thick fog across the Banks of Newfoundland. These banks are thick with trawlers of all nations as well as icebergs and the smaller growlers as they are called, each quite big enough to cut a hole in a yacht along the waterline.

I felt fine, ate well, and never even worried about being seasick. My diet was varied; I enjoyed preparing a meal, and often my mouth would moisten in anticipation while it was cooking.

On arrival I was morally and physically fit. My yacht was in a good seaworthy condition and I felt quite justified and confident about sailing back home again alone. I felt a great sense of humble achievement in having brought the yacht back home again alone without trouble of any kind – achievement in having properly fitted out the yacht for the job as well as actually sailing her.

My round-the-world voyage was an even greater test of whether I could stand doing it on my own. There were even more remonstrances than at the start of the Atlantic crossing, and after the setbacks on the first attempt there was more head-shaking.

Of course, it was a bigger undertaking. The distance was about ten times as great; the periods alone on the ocean were incomparably longer and there were fewer sea birds or dolphins for companionship. It was a venture on a larger scale but I think everything that I have written applies equally to a single-handed circumnavigation of the world. I was alone but not lonely and I had confidence, so that if I worried it was only when I could not get radio contact to let those at home know that I was all right.

Then again, was I frightened? In a sense I was at times. I had to face more storms and more bad weather than in the Atlantic. In parts of the voyage I was out of reach of any outside help – particularly in the Southern Ocean, which was bleak and cold and frequented by few ships or aircraft. There indeed I was alone where major damage, such as dismasting (which on two occasions nearly occurred), could mean literally months of delay, possibly even failure to survive. So in heavy storms and gales I was at times frightened, but it was never extreme or paralysing fear. Whatever the weather, I was always ready and never too frightened to go on deck if the need arose. A better word than 'fear' would be 'anxiety' or 'apprehension' or perhaps 'tenseness'. It made me cautious so that I took the seamanlike action of not carrying too much sail. I felt it would be better to arrive home late than not to arrive home at all.

So you repeat 'Why alone?' I hope you know now, and that these words of mine, inadequate as they are, have conveyed a picture of the single-hander and his mental make-up. One of the great things I have got out of my ventures is the meeting of such a wide circle of real people – people who are *somebody* just because they are sincere, friendly, and helpful. It has been a heart-warming experience and I have made many good friends.

PART 1

I *Early Days*

MY early days were spent in Canterbury where I was born on July 13th, 1908. I was the third child of my parents and had one brother and three sisters. My father was an engineer, but he took up haulage contracting and was principally engaged in bringing consignments of hops to the breweries and fruit and vegetables to Covent Garden. Our family roots and my bringing up were entirely in the countryside and I do not remember any member of the family who was connected with the sea.

I was considered to be rather delicate as a child, so much so that at five years of age my parents dressed me in a sailor suit and had me photographed holding a model of a yacht, because they thought that I might not live to become a real sailor. I had to be wheeled to school in a pushchair and was unable to join in the fun and games. However, I fooled them, for as the years passed I gained in strength. I suppose this handicap really was a help to me in later life because I used to dream of adventure. On winter evenings my sister, Muriel, used to read to me and I would sit on my bed enthralled by the stories of adventure, particularly those about the sea.

My education started at St Paul's primary school, but later on I went to the Simon Langton Grammar School at Canterbury. At that early age I was beginning to be an individualist and wanted to be doing things. As I grew older and stronger I became keen on sport and in my early teens I joined the Canterbury Athletic Club and participated in long-distance races and rowing events. I read still more adventure books and I pictured myself trekking over Africa or exploring up the Amazon or trapping in the frozen lands of Northern Canada. But most of all I wanted to go to sea, to sail to foreign ports in the Far East with romantic-sounding names.

Despite my lack of robustness I had a happy childhood and when I left school at sixteen I obtained a job in an insurance broker's office. During my last days at school and when I had this first job I worked for months on a model of a windjammer. It was three feet long and I worked on the detail of every piece of gear and rigging. I suppose about this time I was dreaming of the Horn of which I had read so much, and certainly the model which I had built added to my dreams. I did not think I was cut out for life on an office stool and I still longed for the sea. One day I made a journey to Gravesend to try to obtain a job on a merchant ship. I had a rude awakening when I was told to go home and grow up.

I was restless and after about eighteen months in the office I left to join my father in his haulage business. I was happier there as my father trained me to overhaul and service heavy lorries; this was to stand me in good stead for the future.

One of my hobbies, besides reading books about the sea, when I was in the insurance broker's office and with my father, was an interest in racing pigeons. I had become quite a knowledgeable fancier and used to dispatch my birds from Canterbury Station for the long-distance pigeon races. I marvelled at their unerring navigation for they raced to the Channel Islands and France as far as Bordeaux and inland to Rennes. San Sebastian in Spain was an even longer flight.

Like the pigeons I seemed to have the travel bug. My brother Dennis was abroad, having a life of adventure on India's North-West Frontier. He was in the Army, so when I was about twenty I decided to emigrate also. I sold my racing pigeons at a good price that gave me enough for a one-way ticket to Canada and enough to start off with. I packed my few belongings and set off for the Prairies.

I soon found work on a farm in West Edmonton, Alberta. I was the only hand employed, and I lived with the family. It was an all-round job as I was a farm hand, cowboy, and labourer. The work was very hard, though, and the first thing I learnt was to round up the horses which had been out all the winter. They were half wild and resented having harnesses put on them. I had eight in a team on a three-furrow plough, and the horses wanted some holding.

Then came the drilling of wheat, and while that was growing we were busy clearing virgin land. I became an expert with an axe, and could split a match on a log.

A particularly arduous job was roadmaking. Farmers were expected to supply a team of four horses and labour as their contribution to making roads in their area. We had to travel ten or fifteen miles to the site that summer. The State Government had the foreman there and implements, in the form of large scoops, to which the horses were hitched. These scoops were dragged along, lifting up the top soil from a hill or mound in the prairie, and up-ended in the hollow farther along, depositing the earth there. The team would then return with the empty scoop and repeat the operation. The man in charge of the team, walking behind the scoop, on rough ground, was very weary by the end of the day, when the horses were hitched back on to our wagon for the long trek home. It was in the heat of a scorching summer, and as the horses were much worried by flies I would hang little pieces of sacking over their nostrils, to help keep off the flies. It was dusk before we got home, and dark by the time the horses were put away and fed. Then we had supper and tumbled into bed, to be ready at dawn for another day. My hands and arms were stiff and sore with holding the horses and I would have to lever my fingers open in the morning. A fresh team of horses was used the next day, but I had to keep going.

With five young children it was hard, too, on the farmer's wife. She was a frail little woman from Scotland. At times when she felt low she would say in her Scottish accent, 'Oh, to be back in the Old Country'; but at other times she would remark, 'This is a great country.'

It was a tough life, but it was the sort of life I wanted, and it toughened me too. However, conditions were so bad at the time that, after about a year, the farmer really couldn't afford to pay me. So, with no prospects in Canada, I returned home to my father's business in Canterbury, no richer in pocket but richer in experience and much fitter in health.

I think that I inherited some of my father's engineering gift and 'know-how' with mechanical things. I worked as a driver and mechanic, and managed to save some money.

I married when I was twenty-three and for the next six years worked with my father. Our first two children were born – both boys – Alec in 1932 and Michael in 1934. Just before the war in 1939 we found a smallholding for sale at Littlebourne, about four miles from Canterbury. I had long felt the desire for a life in the open air, and we took it. It was exciting to start on my own and live in the country.

It was not long before the clouds gathered over the horizon. On the outbreak of war in 1939 I volunteered for the Royal Navy and was called up in 1940. I did my 'square bashing' at Skegness and was drafted to Sheerness in the Thames Estuary, where I worked on the old trawlers being used as minesweepers. It was interesting, if dirty, work on those old coal-burners with triple expansion engines.

I travelled to Liverpool for my first seagoing draft, and found my ship lying in a muddy creek. She was H.M.S. *Leith*, a sloop of about one thousand tons; a comfortable little ship with twin turbine engines, and clean as a new pin. Convoy duties were her job, and we were soon at sea *en route* for Freetown.

This was the first of many convoy runs with her into the Mediterranean, Bathurst, Newfoundland, round the north of Scotland and the East coast. I well remember one attack that was made on the convoy we were escorting. A German bomber from the west coast of France was running in to attack. We let fly with our 4.7 gun on the aft deck and shot his tail off, much to the surprise of the gun crew. We picked up one of the crew, who promptly gave the Nazi salute.

At times escort ships were in extremely short supply, and a convoy of fifty or sixty ships would be escorted by only three small warships. Sometimes attacks lasted days on end, and at other times we would do a whole trip without incident. During an attack by a submarine we would zigzag, trying to pick up an echo on the echo-sounder; if we did we made a run over it and dropped depth charges. These would explode and the impact on the ship was terrific. We, in the engine room, shut down below, were always anxious for information as to what was happening.

The clang of the speed indicator to full speed ahead was the first indication of anything happening. Then the jar up through the footplates to one's ankles as the depth charge exploded would tell us of an attack. Oil and debris on the surface would tell of a 'kill', but not always. Sometimes it was discharged deliberately to put us off. However, we were credited with two submarines while I was with the ship.

One morning, after a bad night during which we had lost several ships, we were ordered to sink a ship that had been torpedoed and was in a crippled state. The crew were Chinese and, in the panic to leave the ship, they tipped the lifeboat up, then dropped it on to men already in the water. The ship was loaded with bales of wood and was a long time sinking. We picked up the crew, and every spare space in the ship was full of Chinese seamen.

In the later part of the war I was mostly on the northern convoys, and it was then that I really got acquainted with the wild North Atlantic ocean. Conditions were tough, with constant alerts and alarm bells ringing, so we never got enough sleep. However, we welcomed bad weather as this increased our chances of getting through undetected.

I remember one terrible storm. We were lying with just enough engine revs. to keep steerage way and *Leith* was rolling monstrously. The sea was a mass of white foam with spray in the air like a mist. The ship was washed from stem to stern and stanchions were bent along the side deck. In the morning our convoy of fifty or so ships were dispersed and only a few were visible. We spent a couple of days rounding them up.

In 1944 I was put forward for a commission and left the ship. I was promoted to Sub-Lieutenant and, after a course on diesel engines, was sent to Warsash and given charge of six landing craft. My nerves were frayed and I couldn't sleep, and one day I collapsed and was taken to Haslar Hospital. From there I was taken to Bristol where I remained several months before being invalided out in 1945, with the rank of Lieutenant R.N.V.R. The period on convoys had taken its toll of my nervous system, and as the doctor said: 'It has taken time to break down the nerve tissues and it will take time to build them up again. Sun and fresh air is the answer.'

Back home I felt like a deserter, and it was some time before I could face people. I threw myself into the work of my market garden and nursery: growing vegetables, tomatoes and flowers for the London markets. Gradually my health was restored, as I was leading an outdoor life and my mind was occupied with things on which I became increasingly keen.

By that time we had a family of four children, as following the two boys we had two daughters ten years later: Anne in 1942 and Jane in 1944.

Meanwhile my father's haulage business was swallowed up by the nationalization of road transport; but my own business prospered.

After a few years I bought a bigger nursery at Ash. This enabled me to grow many more flowers, especially carnations and chrysanthemums, which I was proud of. I think Rose became quite a name for flowers. But financially the move was a big mistake, because the only way I could buy the bigger holding was with the help of a bank loan. Much as I enjoyed the life, it was hard work. I had not enough working capital and skilled labour was difficult to find; after about four years, when I had a bad season I found I could not reduce my overdraft when the bank pressed me. I was forced to sell up.

We took a retail fruit business on the coast at Herne Bay. I was still desperately keen on sailing and read all the accounts of lone voyagers and the yachting magazines. I wanted my own boat. A new one was out of the question, so I bought an ex-ship's lifeboat and spent the next five years converting her. Being on the coast the business was very slack in the winter, so I had time to spend on making a thorough job of the boat.

About this time my marriage began to break up. We separated after twenty-eight years, and I gave up the business. I took to living on the yacht alone and, with Ramsgate as my base, cruised single-handed across the Channel, the North Sea, and the West Country. I was alone with my thoughts and memories.

2 *Taking to Sail*

As *Neptunes Daughter* was my first boat I will add a few details of how she was built. She started her life as a German ship's lifeboat. I bought her at a yard in Southampton and she was hauled to a friend's yard in Broadstairs, just along the coast from Herne Bay.

She was a mere shell with the original buoyancy tanks still in her and the thwarts. I set to work and stripped her right out. The sternpost of a lifeboat is not very heavy, so I put in a new and heavier one to allow for the bore for the stern gland of a propeller shaft, and to carry a larger rudder the yard had supplied me with. *Neptunes Daughter* was built of larch on oak frames and was quite sound. I drew out detailed plans of her proposed construction and layout all to scale. Firstly I considered her bottom planking of $\frac{5}{8}$ in. thickness too thin, in the event of her taking the ground at any time, so I doubled the planking to the turn of the bilge, laying plank on plank in clinker-built fashion as she was built. I put a black bituminous solution between the planks and copper-riveted them up. I then put two heavy oak bilge keels along to stiffen and strengthen her and added six heavy oak floors across inside, dovetailed into the keelson. An iron keel was constructed of two lengths of 'H' section girders bolted together, one on top of the other. In the sections I put scrap iron and filled it with cement. This was bolted on to the wooden keel by heavy 'U' bolts going right through the oak floors. I then boxed the whole lot in with Marine plywood, making a neat job of it.

The next job was to raise the topsides. I raised them com-

13

pletely from stem to stern and put oak frames alongside each
old one to continue it up. I then planked her in clinker fashion
up to deck level. Deck beams were put across and the coamings
built up for the coachroof. These were of one piece cut to shape
to go the whole length of the fore-cabin, the centre cockpit and
aft cabin. In the centre cockpit I installed a Thornycroft
'Handy Billy' engine, with two water tanks placed one each
side. The fore and main cabin was fitted with a dinette one
side and a bunk the other. The bunk had folding sides that could
be opened out to form a double bunk. A 'Taylor' pressure paraffin stove was installed and a plastic sink unit with a fresh-water
pump. Every corner was utilized as cupboard space. The aft
cabin had a bunk but I used it as a bosun's store with a solid
bench and a vice to work on.

I decided to rig her as a wishbone ketch so the masts were the
next job. I visited a ladder maker and selected two poles from
Norway of spruce some 32 feet long. These still had the bark on.
It was a long job cleaning this off and shaving them down for
the mast bands to take the rigging. I gave them several coats of
boiled linseed oil. I then made the rigging up, all myself and to
the lengths shown on my scale drawing. Deep chain plates,
made by myself out of galvanized water-bar, were bolted right
through the stringers. The bowsprit, main boom, and mizzen
boom took a long time too. I cut them and planed them down
from solid spruce and was very pleased with the results. The
wishbone was a problem. It was a long time before I could find
the right 'flow' of the curve. At last I found it when I visited a
yard at Whitstable – Anderson, Rigden, and Perkins. The
manager there was very helpful and found the answer in a text-
book he unearthed. I drew out a scale drawing and he kindly
made it up for me. It was made of laminated spruce, bent and
glued round a jig. I drew out the sail plan and called on young
Mr Goldfinch of Whitstable, a sailmaker in the old tradition,
whose father before him had made the heavy sails for the old
Thames barges. He made up the sails from flax, very tough and
durable. The mainsail went right up through the wishbone to
the top of the mast, giving the effect of a gaff rig with a topsail
attached. The aft end of the wishbone, which went up and down
the mast with the sail, could be controlled by a vang going up

through an eye in the end, to the top of the mizzenmast. This enabled the wishbone to be controlled when lowering the sail, as well as to stop it laying out to leeward too far, when sailing.

The reason why I chose the name *Neptunes Daughter* was because to me the sea is a living thing – Father Neptune. I respect it and pay homage to it. It can be cruel and relentless and will take its toll of those treating it too lightly. As *Neptunes Daughter* I hoped Father Neptune would treat her kindly. Every plank, bolt, and screw I had put into her had been done with the idea of withstanding the stresses and strains that are constantly being exerted in a seaway.

By now she was a heavy boat, and the time came for launching her. I borrowed a trolley from a local boatbuilder, and together we moved her on to it. We enlisted the services of a lorry owner and towed her down through the narrow streets of Broadstairs to Viking Bay and out on the sand at low water, to be floated off with the tide. I stayed on board and as she lifted with the tide I felt tremendously excited and thrilled. The engine started without trouble and I put her alongside the wall to step the masts. She made a little water to start with but the planks soon took up and she was dry after that.

The next day the masts were stepped and I was pleased to find the rigging was all just the right length when the rigging screws were tightened up. After a couple of days putting things shipshape a young boatbuilder friend and I motored along to Ramsgate Harbour where I moored in the inner basin. Mr Goldfinch came down with the sails which were bent on and she looked like a real little ship. I took to living on board and took *Neptunes Daughter* out for short day sails to get the feel of her. She handled well and carried her canvas well.

I had studied coastal navigation, but as I intended sailing deep sea, I needed celestial navigation. I bought an ex-Government sextant. My son, Michael, who was now a deck officer in the Merchant Service, gave me my first lessons whilst on leave, but to get more proficient I wrote to Commander Lund of Torquay and arranged to sail down to see him. I spent a pleasant two weeks with him while he taught me the use of the sextant and how to arrive at one's position from the resultant reading.

I returned to Ramsgate. My son's ship was being refitted in a floating dock at Rotterdam. I thought it would be a good idea to sail over and see him. I left Ramsgate one afternoon and the next night picked up the flashing light at the entrance to the canal. It was a great thrill to make my first foreign landfall, and bang on target too. But I remember a misty rain came down and blotted out everything before I reached it. However, I entered at the Hook safely and tied up by the railway station. The next day I cruised up the canal and tied up alongside the floating dock in Rotterdam. I spent a good week there, during which time I was entertained by the Yard Manager and shown around. I then sailed for Amsterdam. Locking in through the locks at Ijmuiden I lay alongside a moored barge for the night and continued on up the canal the next day. I tied up in the little yacht harbour at Amsterdam on the opposite side of the water to the city. No charge was made, water was laid on, and petrol and paraffin were wheeled round to me with no trouble at all.

I then sailed up round the islands to the north and back down the North Sea to Ramsgate, where I wintered.

The following summer saw me cruising to Calais and Boulogne, and down the channel calling at Weymouth, Torquay, Mevagissey and Penzance and Cherbourg. I had a welcome wherever I went and *Neptunes Daughter* was greatly admired as being a sturdy seaworthy boat. I always left on good terms with the local harbour authorities so that I was always welcome when I returned. I think this is important.

I had met Dorothy, now my wife, while completing *Neptunes Daughter* and the next spring we married and planned a sailing honeymoon. We set off from Ramsgate intending to sail to Spain. Our first stop was Torquay. We had caught some mackerel on the way and intended having them for supper. Dorothy cleaned them in the cockpit and laid them on a plate on the side deck while clearing up, when down swooped some seagulls and, picking up the fish, took them over to the next yacht to eat. We opened a tin for supper. On leaving Torquay we ran into a mild blow, the sea got up and Dorothy felt ill. We went into Plymouth in heavy rain and mist. After a few days' rest we set out again to cross the English Channel. We rounded

Ushant and then headed across the Bay of Biscay. We were about halfway across when a gale sprang up. I shortened sail to storm jib and mizzen, but it increased to severe and I took all sail off her and streamed a sea anchor from the stern.

This was my first experience of riding to a sea anchor and I was amazed at the pull it exerted on the rope. It was bar taut but still the yacht lay with the wind just aft of the beam. I decided to move it, to stream from the bows. I pulled the anchor chain from forward round the shrouds to the cockpit and shackled it on to the wire strop that was the inboard part of the sea-anchor warp. The warp was made fast round a large cleat aft and my idea was to cast this off and take the strain on the chain from forward. I instructed Dorothy to throw the chain over the side when I, waiting for the yacht to slacken the warp, unhitched it from the cleat. I did this and called to her to let the chain go. She held on too long however and the warp, taking up the strain again, dragged the chain over the coaming with her fingers under it. They were cut to the bone and were a ghastly sight.

She promptly fainted. There was I, in a gale, getting her below and tying her hand up. We had a brandy. The yacht was lying quite comfortably and I cooked a hot meal of baked beans and sausages and made a cup of tea. We settled down for the night.

It was a rough-and-tumble experience. The hiss of the Tilley lamp was drowned at times by the whine of the wind and roar of breaking waves, but we were quite snug in our little cabin. Tilley lamps hiss away and give off warmth and comfort. I looked out at frequent intervals during the night but no other lights of shipping were seen. It was pitch dark on deck, with rain; all I could see was a white flash as a wave broke alongside. *Neptunes Daughter* behaved very well though and lay quite dry, riding over the waves beautifully. I was proud of her.

The next day the gale dropped away. I put a clean dressing on Dorothy's hand. We thought it wise to make port and a doctor. I hauled in the sea anchor and headed for Brest. A sun sight put us at about 200 miles south-west. Two days later we picked up a light which I identified as marking the channel and we entered Brest harbour the next day, tying up in a little harbour basin next to the naval harbour. We found a doctor who dressed

Dorothy's hand and gave her an injection against poisoning. He found the wound quite clean and remarked that the sea water had cleaned it. She was very brave over it all and carried on as best she could. We lay at Brest, quite undisturbed for two or three weeks, when we decided to return home. We sailed for Weymouth, which we made a couple of days later, having had a blow off the Shambles and heavy rain. Dorothy's hand was still painful, so she returned home to see her own doctor. The tendons were damaged and she attended hospital for massage and treatment.

Meanwhile it became necessary to earn some more money, and using Weymouth as a base we looked around for a business. Later I sailed up to Poole and stayed there over Christmas. We wanted a business near to the coast and suitable moorings for the yacht. At last we found what we were looking for, in the shape of a good-class fruiterers' in Southsea.

We called and saw the owner, a Mr Dopson, who said that it was practically sold, but that there was no harm in our looking around. We liked the set-up, and Mr Dopson appeared to like us, and when the cat 'Tosh' made a leap at me and curled himself round my shoulders that settled it. Mr Dopson said he would call the other sale off and let us have it. The sale went through. We sailed the yacht to Portsmouth where we moored in Gosport Marina. It was February 1961.

Not much cruising was done during the next two years, for I was busy with the business. In the spring of 1963, however, I was reminded of the second single-handed Transatlantic Race to be held in 1964. I was interested in the first one, which was held in 1960, but *Neptunes Daughter* wasn't the sort of craft in which to attempt the North Atlantic.

However, I was determined to have a go in 1964. I sold *Neptunes Daughter*, but not without a tear being shed. I looked around for a suitable craft. The lightly built slim racers did not interest me. I wanted a strong sturdy craft that would take the pounding I knew she would get.

Of those I looked at, usually one glance was enough to tell me that they were unsuitable. At last I had particulars of a cutter lying at Yarmouth, on the Isle of Wight. She was built of padouk frames planked with $1\frac{3}{8}$ in. teak. I went over and viewed her from the quay. I liked the look of her. I liked her sheer, and her buoyant-looking stern. I met her owner, a Mr Cambridge, who had built her himself in Calcutta. I went on board and was impressed by her solid strength. Below, this was evident in her heavy timbers of padouk, a wood harder than teak, as the shipwrights discovered when drilling holes through it. She was rather old-fashioned in her rig, though, and I knew I should have to change that. I bought her after a survey had pronounced her sound. Her name was *Lively Lady*.

I went to see Captain John Illingworth, of Illingworth and Primrose, and he drew me out a new sail plan. The mast, a new metal one by Proctor, was moved slightly aft. The bowsprit was shortened, as was the boom. She was flush decked, so we added a low doghouse where the hatch was. I sailed her round to Mike Attrill's yard at St Helens (I.o.W.) for this work to be done.

I also called on Colonel (Blondie) Hasler, who had designed a self-steering gear, which was highly thought of. He came over to see the yacht and drew out plans for the gear to be fitted.

The sails were made by Lucas of Portsmouth.

By February 1964 we were ready for trials, and Sir David Mackworth, who was then working with Illingworth and Primrose, came over with me. It was a great thrill to stand on her deck and to feel her alive at sea. She went well – but one or two adjustments had to be made to sheet leads and other things. After this had been attended to we sailed her to my moorings in Langstone Harbour. One or two short sails were carried out to get the feel of her, but I was too busy to go any distance.

'Blondie' Hasler expressed a wish to come out for a short sail to test the steering gear. He came over with his assistant 'Jock' McCloud and we took her out into the Channel. The self-steering gear worked well on all points of sailing.

I was ready to start storing up with provisions and requirements for the race. People were very kind and many gifts of food and stores were made to me to help me on my way. A party was held at the Eastney Cruising Association to wish me luck, and one Sunday in the middle of May I dropped my moorings in Langstone Harbour and sailed out, to the cheers of club members and escorted by numerous small craft. I was on my way to Plymouth and the start of the race.

R—C

3 *The Transatlantic Race*

'**D**ID you ever see so many boats in Plymouth Sound?' was the question put to me by Theo Mashford, of Mashford Bros. shipyard, and I agreed – I hadn't.

As the gun went for the start of the second single-handed Transatlantic Race on May 23rd, 1964, the fourteen contestants headed for the line, together with literally hundreds of small craft of all sizes and shapes carrying interested viewers. There were in fact some near misses and Val Howells in *Akka* wasn't missed and had to put back for repairs. Some of the fourteen had their wives seeing them off, some had sweethearts, and some had both, though not on the same craft.

The 3,000 miles ahead of me didn't worry me so much as the thought of ramming and sinking one of my rivals, and so I decided beforehand to lay back and let those who wanted to contest the honour of being first across the starting line. This is the one race, I thought, where minutes would not count at the start, though I thought the next race might well be different.

It was only a gentle northerly breeze and we weren't exactly tearing along towards the breakwater. I couldn't make out the leader amongst the crowd, but I did see the Frenchman, Eric Tabarly, setting the pattern of his determination by breaking out his spinnaker. By evening everybody seemed well strung out and I seemed to be bringing up the rear. However, I had a good meal and generally tried to get the yacht shipshape for seagoing. I was a little weary, with all the excitement and the press of last-minute jobs before the start, to say nothing of the several parties we attended. In fact, our well-wishers and friends were our worst enemies in their desire to give us a good send-off. The quiet start was therefore, to me, not unwelcome.

The next day, Sunday, found me becalmed approaching Land's End and I started to work out that it would take me at least six months to cross the Atlantic at this rate.

Lively Lady was then a cutter. She was very heavily built and not a greyhound by any means. With a good long keel, good sheer and freeboard, and buoyant ends, she is, however, a very safe seaboat. She proved this to me time and again in the weeks to follow. In light airs she is slow, and now I realized I was a little down by the bows. I had received much in the way of gifts in stores and in the last rush this had all been dumped down the forehatch. I moved some aft and then started ditching anything I could lay my hands on that was not of value, including an old 50-lb. anchor of unknown design. I must admit to being a great hoarder, until at last I find the ship full of useless bits of gear, and so on.

Looking around I saw *Jester* way in under the cliffs and Blondie Hasler told me later that he had to scull to get away. With the approach of darkness, mist came down – but also a breeze got up, and I was able to pass the Seven Stones Light Vessel (hearing her but not seeing her light), clear the Scillies to the north – and head west, south of Ireland. My intention was to go the Northern route and I had marked the Great Circle route on the chart with the idea that when I couldn't hold this course, I would go north. However, it was a fresh northerly wind and we made good progress – in fact, on Monday I had to reef the mainsail and set storm jib and spitfire staysail. I had a good spray-hood which was a joy and comfort and I could sit in the open hatch and watch from there. I was doing just that when I was startled to see *Jester* crossing my stern heading more northerly than I was. She seemed to be tearing along and was a great sight.

The wind continued kind in direction, but sometimes was very light, for several days, but then a short but fierce southerly gale blew up and it was only after difficulty that I got the big genoa down safely. I carried the reefed main and storm jib all right and kept going more or less in the right direction. A couple of days later I sighted *Jester* once more astern, which

gave me incentive to change headsails to pull away and lose
him in the mist. It just proves one needs spurring on to race
full out. I was too complacent with my comfortable progress.

May 29th was awful, with sheets of solid rain and a south-
east gale. I held on to the reefed mainsail too long, but got it
down all right and lay a-hull. The *City of Bristol* circled me and
flashed, but as she was continually hidden from view by the
huge swell I was unable to make it out and we gave it up. I
gave her a confident wave, she gave me three toots and she was
away. He reported me to Lloyd's, however, and this was the
last my wife heard about me until June 29th. The seas were
awful, and yet impressive, and I watched from the cockpit the
huge Atlantic rollers marching up with their white breaking
crests, showing blue water through the steep tops and some
hitting very hard. The motion was severe and anything not
stowed properly below shot all over the place. My plastic jar
of white sugar leaped off the shelf, came right across the cabin
by the galley and distributed its contents over everything,
including the charts. It took me days to get rid of wet sticky
patches.

During the night I wrote in my log, 'Very rough night.
Stiff and sore all over with the violent motion. Thankful to lie
in bunk. Must remember that the single-hander must conserve
his strength. Whose idea was this race? An idiot's?'

The next day it eased, but as always a gale leaves a swell and
lumpy sea. I prepared a nice salad lunch of ham, boiled egg,
lettuce, and tomato, with salad cream and bread and butter.
As I turned round a moment to put the kettle on, it was
deposited on the cabin sole. What did I do? Picked it up and
ate it of course.

So the days went by. I ate well. I always start the day off at
dawn with a cup of tea laced with whisky, and put the porridge
on. I enjoyed this with plenty of brown sugar, a handful of
raisins, and milk. A little later I would fry up eggs with the
remains of yesterday's boiled potatoes, or sausages, or baked
beans and fried bread – all washed down with a hot coffee.
After the routine jobs of the ship such as filling oil lamps, and
taking a sight if possible, I would have a mug of hot cocoa
and cake or biscuits. If conditions permitted I would try to

get a couple of hours' sleep, after which another nice cup of tea with cake. In the late evening I would get a good meal, starting with soup and a fry-up of a pan full of onions and sausages or cold tinned meat or salmon and fresh lettuce or tomatoes. I also made some stews of tinned steak, butter beans, mixed vegetables, and boiled potatoes, and I generally added some little suet dumplings I made – not forgetting a pinch of mixed herbs. Granary bread was given to us at the start and kept well throughout the voyage. I hung it up in net sacks and although it went green on the outside the middle was good. I had a supply of oranges, apples, bananas and lemons, and after every full meal I had one of each, so you see I lived well. In the dark hours, or at any other time I felt the need, I also had my special pick-me-up of squeezed lemon, two or three large spoonsful of honey with hot water, added to which was a generous splash of whisky. Guaranteed to warm up your tummy. My doctor also gave me a bottle of vitamin pills to take every day – which I did. As first-aid equipment he also gave me pills and ointments for every eventuality – including splints for broken arms – none of which, I am glad to say, were needed. At no time did I even feel like being seasick.

In severe conditions, when things were too hectic to cook, or I was too tired, I had some self-heating tins of Horlicks Malted Milk – as supplied to lifeboatmen. This was good – but I found I needed a good dash of brandy to enjoy it.

My days were full. There is always a job to be done of some kind and, far from being bored, I enjoyed every minute, and I didn't seem to have time to read books.

One morning I was aroused from a sleep by a long moan or sigh, and on going on deck saw a huge whale – much longer than the yacht – right alongside. He dived and a few minutes later surfaced on the other side and repeated the sound. He was, of course, 'blowing' or breathing. Then I found there were two and they appeared to be investigating this strange 'fish' sailing along in their territory. I wondered what would happen if they surfaced under the yacht. I admired and respected their great power and grace in the water, but was relieved when they departed.

Another time I was awakened by a breaking of waves and

yet the yacht was quite steady in motion. I found a huge company of dolphins around me. Literally thousands of them, like a huge flock of sheep, leaping and diving in such a happy manner, as far as I could see. Big ones and baby ones all keeping station with me for about two hours. They were good company and I shouted with joy to welcome them.

The morning of June 13th I was startled to see a sail astern. Who could it be? It could only be one of us, I thought. However, he gradually crept up on me, and then I made out his number – 4, which I knew to be Bill Howell in *Stardrift*. As he drew near we hailed each other. A Russian trawler on the way home circled us, her crew lined up agape, evidently amazed at seeing two British yachts in convoy in mid-Atlantic. *Stardrift* then made off on a more southerly course than I was pursuing. About this time I found that I wasn't enjoying my cup of tea and for several days tried to discover why. I suspected the tea, the teapot, the kettle, until at last I discovered it was the water. It was almost like sea-water. Evidently with the weight of water on deck at times, sea-water had syphoned into the tanks through the little overflow goosenecks which come out on deck. However, I always carry a reserve supply of a couple of two-gallon plastic cans of water for just this sort of emergency, and these saw me through for drinking. The other was all right for cooking.

I was now approaching the Grand Banks of Newfoundland, where trawlers of all nations fish, where ice is floating around and where there is fog nearly always. I didn't see any ice, though St John's radio gave repeated warnings of it. The reason, of course, was the fog. Thick swirling, dripping, freezing cold fog. Yet we were sailing well at five or six knots. At times I couldn't see farther than the bows and it was a little unnerving to see that, however good one's lookout, any object would be struck as soon as seen. The only thing to do was to say a prayer, and trust in God. Which I did. At night the blackness could be felt and on shining a torch from the comfort of the hatchway under the spray-hood, all one could see was swirling fog round the wind vane of the self-steering gear. Often foghorns sounded very close, and my blood froze in my veins as I strained my ears and eyes in their direction. I think my radar reflector was doing its job and they picked me up on their screens, as they generally stopped blowing as I got abeam of them and passed. I had this, off and on, for the last 1,000 miles – right down past Sable Island, that graveyard of wrecks which I went north of, and off the Nova Scotian coast, which I got too close to, round Nantucket Light Vessel, and up to the Brenton Reef Tower – the finish. The fog and calms were the most frustrating of all, when one is so near the finish, and it took me nearly three days to do the last 100 miles. At one time I thought the passage could be done in about thirty days – but it was not to be. It was 36 days.

After crossing the line in a near calm I hove-to till daylight. I then noticed a pilot boat dropping a pilot to a big tanker astern of me. Hoping he would contact me, I got my long warp out to pass to him. He did just this and I was ready. On asking him where he would drop me, he replied, 'I will take you where all the others are', which depressed me as I thought, 'I am really last after all.' But I didn't let on to him and just replied, 'That's fine'.

After he cast me off and I drew alongside the Port o' Call Marina Pier, Peter Dunning, an Englishman who runs it, and other willing hands secured *Lively Lady*. Others brought me coffee. Carol Heinz of the Canadian Broadcasting Co. asked for my home address and rushed off to cable my wife, and it wasn't until then that I was told I was fourth in. That was the pattern of the welcome we received. Smiles and handshakes, always with the request, 'Anything we can do?' It was terrific and overwhelming. The Mayor had us to a civic reception at which we were presented with a copy of the Council Resolution acknowledging and recording our 'feat in crossing the broad expanse of the Atlantic Ocean'. The local radio and news reporters were helpful and kind without being pressing. The whole effort of the race was well worthwhile, if only to meet these wonderful people. I was proud to be flying the Blue Ensign. People would walk down to see us and were anxious and proud to make it known if they had British ancestry. I could fill pages with the story of the many kindnesses received right up to the time I left, when a gathering of local officials and friends came down to give me a great send-off home. As

well as wishing me safe return journey they all demanded that
I come back some time. Francis Chichester and his wife were
there and David Lewis and Mike Ellison escorted me out and
I was on my way.

The homeward journey followed much the same pattern. As
I passed the Brenton Reef Tower, *Sovereign*, the America's Cup
challenger, was out sailing on trials with *Norsaga* and I signalled
'Good luck' to Peter Scott and his crew, which they acknowledged.
Again the fog closed in as I was near the Nantucket Light Vessel –
but I passed north of her this time without seeing her. The big
ships keep south but plenty of trawlers were all around as well
as coasters, and their higher-pitched sirens mingled with the
deep-throated notes of the big ships, like a lot of tom cats,
screaming 'Get off my pitch'. Once I heard one ship sound
four blasts in quick succession in urgent warning to another
in danger of getting too close.

'It's a downhill run home,' everyone said, 'just hoist your
twin headsails and let her go.' Well, for three weekends
running I had south or south-east gales, one of which was
severe, ending in a stinker of a north-wester for the fourth
weekend. Extracts from my log read:

1st Aug. Been a wild rough day and just been able to snatch
something to eat.
2nd Aug. (dawn). The yacht doing everything except stand on her
head and I am bruised and sore all over with being thrown
about so. Should shorten sail really but heading in the right direction under reefed main and storm jib. True Atlantic morning with
overcast grey sky and breaking crests all around. Glass still falling.
(Noon) Lowered mainsail – waves clean over the yacht, wind
shrieking.
(1300) The knife drawer slams shut, crushing and splitting my
finger, makes me feel sick as I hold it in cold water.
(1400) Conditions worse, have to go on to foredeck to lower storm
jib. Wind tearing and screeching at everything, heavy rain,
waves over the ship. Pretty perilous up in the bows – trying to
get a madly flapping sail down – with the bows lifting 20 feet in
the air, and you see a deep valley of water into which you are
dropping. Coming at you is the next huge wave – ready to swamp
you and carry you off as his. You look at the sky and it seems to
be revolving and swaying as the yacht is twisted and tossed

about. Blood from my split finger is all over the sail. Down below everything is wet as wave after wave sweeps over. Cascades of water get under the skylight cover, forces itself through the hinges and sprays the cabin. Damp clothes hung up, get wetter as do the bunks. Decide to have a self-heating tin of Horlicks laced with brandy. My head is in a whirl with the dizzy antics, my eyes ache, my nerves are tense with hearing a hissing monster approaching and waiting for him to strike – Bang – and the yacht quivers and shudders and reels to the blow and sometimes before she can recover another strikes her – smothering her in spray and water. Jars and tins leap in the air on their shelves and a stopper will spring out of its bottle and fly across the cabin. The high-pitched whine of the wind and the shriek in the rigging is getting on my nerves. Had it for 36 hours and no sign of letting up. Have only to say aloud, 'I think it's a little easier', and the elements would scream back at me for daring to suggest they were tiring. Must try to get a sleep. One is entirely on one's own out here so must guard against exhaustion and lethargy.

(*1900*). Decide to rest below. Ought to get some food inside me so make a stew.

(*2200*). Wind easing, so after a hot lemon, honey, and whisky, lay down under the blankets with a cellophane sheet on top. Must have drifted a lot and no idea as to position except that I am mid-Atlantic.

That was what I wrote at the time.

For the rest it was rough, unsettled weather all the time, and again on Sunday the 16th it got steadily worse, though the wind didn't seem to know which direction to come from. On the 17th it was blowing a gale from the west and we were running under the storm jib alone. The wind veered to north-west and as darkness fell we were tearing along hardly able to carry the jib. I wrote:

What a terrible black stormy night it is. God help us. I suppose I shouldn't worry – but I'm a little anxious.

18th. Wild rough morning as dawn breaks and during the morning it seemed to get worse, with heavy rain and visibility down to perhaps $\frac{1}{2}$ mile at times. Haven't had a sun sight for several days, but we should be approaching the Scillies according to my dead reckoning.

I scanned the horizon ahead – but I worked out that with

luck the Bishop Rock Light would appear that evening. It was not certain, however, for as I said, with no sight for fixing and these severe conditions one could easily be enough miles out to be in trouble.

At 1800 through mist and murk I saw a large three-funnel ship away on the starboard bows, going roughly the same direction as I was – but I soon lost her again. This relieved me, however, as I knew she would be going south of Bishop Rock and I therefore felt more confident. Just after that the sky brightened and visibility improved to the north-east and there on the port bows was this tall lighthouse – which could only be the Bishop Rock. Almost as I looked again it disappeared – but I knew where it was, and it was just where I wanted it and to windward. The shriek of the wind and the rough seas sent the yacht crazy and it was time to get the storm jib down before dark. I crawled forward after easing the sheet and got it down. I was lashing it to the rail when out of the corner of my eye I saw a black cylindrical object, with guy wires from it, in the sea to port. It startled me as I thought, 'I could have hit it', but it was to windward and seemed to be fixed and not rolling as a buoy would.

Thinking I was safe, I carried on and got back to the cockpit, but to my horror this object came closer. It was now dusk and it carried no light. I thought I was being carried on to it. I got really worried – something had to be done quickly – and the only thing was to get a baby sail up. I went forward again to get the spitfire staysail up. The yacht did her best to throw me off, and several times I was pitched across the deck and generally treated in a very rough manner. All the time I was looking again and again at this object as I unlashed the sail – when, before I could do more, the thing surfaced. It was a submarine which had evidently been watching my antics and probably thinking what a fool I was to be there at all. This was a great relief as I assumed she could keep clear of me. Hooded figures appeared in the conning tower as she cruised around me and blew her siren, evidently wondering if I was in trouble. I waved them away, however, and the last I saw was when she was submerging again evidently thinking it was a lot more comfortable on the sea-bed.

By this time the Bishop Rock Light shone out of the gloom and I hung my Tilley storm lantern in the rigging and retired to comparative comfort below. It was a wretched night with seas hitting us so hard I really thought something had to give. However, the morning came, the seas went down and I had quite a comfortable sail up to Plymouth. I arrived feeling very battered and bruised, tired and weary. Much more so than on the outward journey.

The yacht stood the double crossing as I expected her to, and my only damage was a broken booming-out pole, caused through holding on to the sail too long. 'Blondie' Hasler's self-steering device was as near perfect as is yet known, and it would have been impossible to undertake a venture such as this without it. Nowhere in the world – including America – is there anything like it, let alone to rival it, and I wouldn't even go cruising without it now.

I wouldn't have missed this trip for anything, and I am extremely grateful to the fates for giving me the opportunity. From the time I met Captain Shaw and officials and members of the Royal Western Yacht Club, to the time I got back here again, I received nothing but kindness, and I only hope I may meet all those good people again.

4 *Planning the Great Adventure*

A FTER the Transatlantic Race I was occupied in writing articles and giving talks to various clubs and organizations. Not much sailing was done, I'm afraid. Short day sails were taken but that was all.

By the winter of 1965–6 I was thinking of what I could do next in the way of a long single-handed voyage. In the back of my mind of course was this desire to sail right round the world, and I wondered if I could do it. I pondered over it. I took stock of the yacht. She was sound enough. The sails would need looking at, of course, and certain items of rigging renewed. I became excited at the thought of the greatest adventure of my life. My wife Dorothy was very efficient; I knew she could manage the business and I was lucky to have reliable staff to support her. Then Francis Chichester announced his plans for sailing to Australia along the clipper route, and the idea came to me to follow him, and to make a match of it. My son had married an Australian girl and was now living there. It would give me an opportunity of seeing him again, and his wife and two children whom I had never seen.

I went back to John Illingworth for advice on putting a mizzenmast on *Lively Lady*. With the shortened boom there was a long counter aft and I had often toyed with the idea of putting a mizzenmast there, to carry a staysail. A mizzen sail aft was out of the question as this would interfere with the self-steering vane, but I thought a mizzen staysail to use when off the wind would help a lot. He and Angus Primrose agreed, and with Sir David Mackworth we drew up a new sail plan.

We visited Ian Proctor's works at Swanwick and ordered a small mast that would serve as a mizzen.

I kept this plan of mine very much to myself, but in spite of this, news got around. I thought it only courteous to tell Francis Chichester of my plan and wanted to be the first to do so, but when I rang him up he told me he had heard about it. His date for leaving Plymouth was August 27th and he suggested I be there to leave with him. I considered this too late for me, however, as he could give me at least three weeks' start, and I proposed leaving Portsmouth on August 7th. The problem for me was to get to Australia early enough to enable me to leave again – after a short stay – in time to get round Cape Horn before the bad weather set in. Boat for boat *Lively Lady* was no match for *Gipsy Moth IV*, which was brand new, designed and built especially for the job, and much longer on the waterline. However, the news became public and the Press announced that I was to race Francis Chichester round the world, although this was not quite true as I had no illusions about the relative speed of the two yachts. Then Group Captain Searle, a yachting journalist who runs a sailing school, suggested to the Lord Mayor of Portsmouth (Councillor C. Worley) that a few citizen yachtsmen might like to get together to sponsor me in another faster yacht. The Lord Mayor called a meeting to which he invited a dozen or so local businessmen. I was against the whole idea. I preferred to sail *Lively Lady*. I knew her and had confidence in her as a seaboat. I didn't want to be sponsored; I wanted to be captain of my own ship and be beholden to no-one.

However, the feeling was that they would like to help me with expenses of fitting out and with no strings attached. I was grateful to them for this spontaneous gesture, it made things much easier for me. I was able to order a new mainsail from Lucas as well as a new large genoa, a working jib, and of course the mizzen staysail.

I spent every spare moment on the yacht doing odd jobs. She was hauled out and we rubbed her down and painted her, varnished the coamings and skylight, etc. The skylight, although watertight in normal conditions, leaked when under pressure from a wave coming aboard, so I screwed it down on to a liquid rubber solution. A friend came and put it on for me. This was also put round the forehatch and proved very effective. I took out the cabin table to save weight and make more room to get through the cabin to the forepeak. I also put the bunk mattresses ashore, leaving only one for my own use.

I have a 'Tiny Tim' charging engine slung in gimbals and

The Cabin

this was taken out and overhauled and new batteries installed. The main engine also was overhauled. As the propeller was pitted I had a new two-bladed one fitted, and marked the shaft so that I could have it up and down behind the sternpost when sailing.

I have always been against having a radio transmitter, on the grounds that if no contact was made people ashore would worry. However, I felt that on a voyage like this a transmitter would be useful to contact points *en route*. I had a 'Sailor' radio-telephone installed and this proved its worth, enabling me to pass messages back home and to receive. This was a normal 'Off the shelf' model; the makers reckoned about 300 miles the limit to transmit, but in the event I was talking to Cape Town at 850 miles and to New Zealand at 1,200 miles.

The new mizzenmast arrived from Proctor's and we stepped that and set it up just aft of the rudder post. I suppose this turns her into a yawl though without a mizzen sail. The self-steering gear was checked over and new tiller lines connected.

In all this work I was helped and advised by my friend Sir David Mackworth and I want to express my appreciation to him.

The Dunlop self-inflatable life raft was sent through Moody's of Swanwick Shore to be serviced. Moody's were very helpful indeed to me and offered me every assistance that I needed, both before I left and should I get into trouble on the voyage. Colonel 'Blondie' Hasler and his assistant Jock McCloud came over to check over the self-steering gear, and suggested some spares to take. Only spare servo blades were needed on the voyage.

During this time we were collecting together stores. Many firms were very helpful over this and sent me free gifts of their goods. These ranged from whisky (given by Brickwoods, the local brewers) to Marvel, baked beans, chicken supreme, and tinned cake. In addition lady friends and relations made me several rich fruit cakes; these keep well and are very nourishing. The local dairy also gave me bottled sterilized milk which will keep for months unopened.

Dorothy, helped by our friend Mrs Baba Sparkes, spent many hours packing these stores in cellophane bags, stowing them in lockers and elsewhere, and making a detailed list.

Charts were a big item. We paid a visit to the Admiralty Chart Office in Portsmouth Dockyard, and received the utmost help and advice. Large scale-charts were obtained of the English Channel, as well as any points *en route* at which I might have to put into, such as the Canary Islands, Cape Town, the South Coast of Australia, and of course Bass Straits, Hobson's Bay, and Melbourne itself. For the return voyage large-scale charts were bought of the west coast of Tasmania, Southern

New Zealand, and then across the Southern Ocean to the
west coast of South America, Chile, Tierra del Fuego, the
Straits of Magellan, and Cape Horn itself. This constituted
quite a pile of charts, and together with the Admiralty list of
lights, Radio Signals, Sailing Directions, and D.F. Beacons,
made a lot of weight.

The Galley

The week before I was due to leave was wretched, wet and
windy with gales blowing most of the time. It was miserable
tightening up rigging screws and shackling on sails in oilskins
and in heavy rain.

My safety precautions were simple and conventional. As on

the Transatlantic Race I had safety harness and lifeline for working on deck and a life jacket. The former I always used in hazardous conditions, but the latter I never wore because it is cumbersome and if one goes overboard when single-handed there is little hope of getting back except by means of a lifeline.

I made a last minute check-up of the radio-telephone and topped up the petrol and paraffin tanks. I filled the fresh-water tanks which were the most important thing of all. I carried about 70 gallons of water in tanks and 30 gallons in plastic jerry cans, which was enough at half a gallon a day to last 200 days at sea; with care it would last a good deal longer.

For the petrol-paraffin engine I took 30 gallons of paraffin, which gives a cruising range of about 150 miles under auxiliary power, apart from petrol which could also be used for cruising as well as for starting. The petrol tank had a capacity of 6 gallons and in reserve I had five two-gallon cans, giving a total of 16 gallons. The 'Tiny Tim' which charged the batteries used only about 3 pints an hour.

Another thing I checked over was my list of tools. I always carry a full set of tools as these are vital for an ocean-cruising man. A single-hander in particular must be able to carry out repairs at sea with his own hands, whether to hull, mast, or rigging. A vice is one of the most essential of the tools. I clamped mine to one of the companion steps when in use and I could not do without it.

Several parties were given me as a 'send-off'. One very pleasant evening was spent with friends at the Guildhall with the Lord Mayor of Portsmouth, who kindly entertained us; and another with my club the Eastney Cruising Association, who presented me with a camera. By the morning of Sunday, August 7th, I was tired and tense. The plan was for me to sail from Portsmouth Harbour to Langstone Harbour where the Eastney Cruising Association had arranged a send-off. I was then to sail back to the Royal Albert Yacht Club where the official start would take place.

5 *A Chapter of Disasters*

I T was raining hard as I left my moorings in Portsmouth Harbour, and sailed out past the Forts and into Langstone Harbour. I tied up alongside the Commodore's yacht, and went ashore to land on the beach.

In the Clubhouse toasts were drunk, short speeches made and I was ready to re-embark. Back on board we cast off and, in turning in the narrow channel, *Lively Lady* just touched the ground. I accepted the offer of the local coastguard launch to pull her head round, and we were free.

The engine was not behaving itself, and was difficult to start. The wind was fluky, and I was in a hurry to get back to the Royal Albert Yacht Club, where another reception had been arranged for me. Coming up to the mooring buoy, the engine flatly refused to start, and the batteries were run down trying. We passed a rope to one of the club members in a dinghy, who made us fast to the buoy. The Lord Mayor, Councillor C. Worley, was on hand with his outboard dinghy to take us ashore, where we were greeted by Commodore Roland-Phillips and a large gathering of members and friends. Many members of the Portsmouth County Club, of which I was Vice-President, were also there to see me off. Another toast was drunk, I said goodbye to my wife and family and I was ready to get back on board. Commodore Roland-Phillips escorted me out to the yacht from the beach. I got under way and crossed the starting line of the R.A.Y.C., at about 3 P.M., under mainsail and working jib, to the boom of the starting gun, the cheers of onlookers, and the hoots of numerous small boats clustered round me.

I passed out through the Forts, set the genoa staysail and headed for Bembridge Ledge buoy. One by one the escorting craft left me, and I was alone.

I felt depressed. I felt I was not ready. Down below nothing was stowed properly, and now the engine would not run to charge my batteries should the 'Tiny Tim' fail. I should have delayed my departure to have got this seen to. But when people kindly arrange a send-off and it is announced in the Press one feels bound to go on.

By 7 P.M. the wind had increased to Force 6, and I lowered the staysail and reefed the mainsail. An entry in my log reads: 'Never have I started with the ship in such a muddle and so unprepared. Even the engine won't run.'

It was slow going, with the wind dead ahead and a short, choppy sea. It was tack and tack next day and it was midnight before Portland Bill was abeam, but with the tide against us no headway was made until early morning.

Niton Radio broadcast a gale warning at 0830 and it began to blow up. I changed to storm jib and put another roll in the mainsail. It was rough going as *Lively Lady* headed into the seas across Lyme Bay; it was raining hard and water and spray were thrown all over the boat as she plunged into the head seas. As the gale increased and backed to S.S.W. I had visions of being driven into the bay. More than one good ship in the past has suffered that fate. It was a rough night. I hove-to, heading south to south-east, to get sea room and make sure of clearing Start Point. Visibility was very poor and, with constant watch for ships, I was beginning to feel very tired. No lights were visible.

Early next morning conditions eased and we got going with working jib, staysail, and reefed mainsail. It was then that I found the steering blade of the self-steering gear had snapped off short. I had a spare one, however, and I got the gear inboard and changed it.

I sighted land to the west and it could only be Start Point; but it was tack and tack to clear the point, and it was evening before it was abeam. Then fog came down and Start Point foghorn boomed out, as well as the sirens of several ships in the vicinity. It was constant watch; but, just before dawn, Plymouth breakwater light flashed out.

I decided to go into Plymouth to get the engine looked at, and also to replace the broken steering blade. I headed for the light, but by dawn it was blotted out by misty rain. 'Typical Plymouth weather', I wrote in my log. I entered the Sound and sailed round Drake's Island. From there to Mashford's Yard at Cremyll it was a headwind and with the wind blanketed by the high land slow progress was made. Finally a motor launch came along and offered me a tow, which I readily accepted, as I was too close to some jagged-looking rocks. After picking up a mooring a boat came out and I went ashore to be greeted by Mr Sidney Mashford, whose first words were, 'Come on in and have some breakfast and a hot bath.'

This was very acceptable, and I was grateful. My top clothes were dried. The oilskin suit I had been wearing constantly had sweated inside and I was as wet inside as out. It had taken me three and a half days to get to Plymouth, against head winds and a full gale.

In no time the engineer was out to look at the engine. After tracing the trouble to the magneto he took it off and found it in very poor condition. It obviously had not been looked at in Portsmouth. I had particularly asked that it be checked over, and had been assured that it was all right. It was returned to the Service Depot for a replacement. This took a couple of days.

In the meantime two new spare steering blades were made and I busied myself in odd jobs round the yacht. The radio was checked over, as it had cut out on occasions on the way down. The expert found a 'dry' connection, and put it right.

It was three days before I was ready to sail again, and I felt rested and more ready. It was the weekend and Dorothy came down to join me. On the Sunday some of the staff from our shop at Southsea drove down and we had a pleasant dinner party at the local hotel. It was Monday before I dropped my mooring at Mashford's and headed for the breakwater.

Outside the breakwater the wind dropped away to nothing, and I lay wallowing in the swell. I lowered the jib and un-shackled the halyard, only to see it run up to the masthead.

'What now?' I thought. There were several fishing boats about, and I thought of asking one of the crew to come and hoist me up the mast. But they were not close enough, so I started up the engine and ran back to Mashford's, where willing hands retrieved the halyard.

Again I set out. 'Third time lucky,' I thought; but it was not to be, as I will tell you. The next day was sunny, the sea calm and a light breeze from the south-west, as I headed out past the Eddystone Lighthouse.

By midnight I was well down channel towards Ushant. The wind fell light and I lowered the genoa staysail, as it was flopping about. We ghosted along on a calm sea, with hardly steerage way. Early in the morning, at about 0330 when it was still dark, I saw a large vessel's masthead lights appear, and I could see the red and green too. I flashed my torch – a powerful one – but she kept coming. I had not enough way on to turn quickly, and there was not time to start the engine. On she came towards my starboard bows. Then we struck. She caught me a glancing blow. Her port bows struck my bowsprit, twisting me round so that my port side scraped along her side. It seemed an eternity as we bumped and scraped along, my shrouds rubbing rusty green paint off her and covering my deck. As she passed, her great propeller threshed the water and sent spray flying over me. Then she vanished in the darkness, with just her stern light showing.

I was shaken, but a quick look round below revealed no leaks. I went forward and found the short, heavy bowsprit splintered. Two of the guard rail stanchions were broken off. I could see the port spreader was bent back; as was the mizzen port spreader. I was sick with worry and despair. I could not possibly go on in this condition. If I had been well on my way it would have been different; I would have had to make do and carry on. But to leave the English Channel in such a condition for a world voyage was out of the question.

When daylight came I bound a rope round the splintered bowsprit and drove a wedge in to tighten it. The pulpit was bent and the port topmast shroud was coated in green paint. I could see a dent in the metal mast where the spreader connected, caused by its being bent back.

I turned the yacht round and headed back to the Eddystone and Plymouth. I was full of misery and despair. What could I say to all those good people in Portsmouth who had given me such a wonderful send-off?

The wind remained light, and it was the next morning before I picked up a mooring off Mashford's Yard at Cremyll.

A boat was soon out and the bowsprit measured up for a new one. The mast was inspected, and it was decided to contact the makers – Ian Proctor of Swanwick – for advice. Their verdict was that the mast should come out of the boat so that the dent could be hammered out. A stainless steel band was then to be riveted round the mast, to stiffen it. They immediately made this band up and put it on passenger train for Plymouth, together with rivets, etc. I telephoned Dorothy, with the news. That afternoon we moved the yacht alongside the wall, under the crane, prior to lifting the mast out.

The next morning I heard from Dorothy that she was coming down. Sir David Mackworth also sent a telegram that he was on his way. It was a great comfort to have such support in times like this.

We lifted out the mast and in the afternoon I met Dorothy and David at the station, and also picked up the parcel from Proctor's. By this time the new bowsprit was made and fitted. The work on the mast was carried out personally by Mr Sidney Mashford and his son-in-law, Peter Lavers, and assisted by Sir David Mackworth. A wonderful job was made of it, too, and the general verdict was that it was stronger than before. The guard rail stanchions were welded on and, on Monday evening, we were ready to re-step the mast on Tuesday morning; I could be away again on Tuesday afternoon, the 22nd August, over a fortnight since leaving Portsmouth. Quite late enough, but not too late, I felt.

We were sleeping ashore at a small guest-house, and the next morning packed our bags and prepared to leave. David had walked down to the yard before breakfast, and we were waiting for him to return. When he came back he brought terrible news. *Lively Lady* had fallen away from the wall at low water, and crashed heavily on to her starboard side. Mr Mashford had heard the crash from his house.

I was stunned and shaken. This was the last straw. It seemed that from the start Fate had taken a hand to prevent my going, and when I survived the previous mishaps she was determined to stop me by pushing the yacht over. As we set off to walk down to the yard I felt we were rather like mourners going to view the body.

On arrival we found sail bags floating and some pig iron laid along the inboard side deck had fallen across the deck and broken the guard rail stanchions on the starboard side.

Examination showed three or four cracked timbers. The caulking was shaken, of course, and all the seams showed up along the starboard side. How she came to fall over remains a mystery. She had taken the ground with the tide for three days with no trouble. One theory is that a tug had passed up the river, sending her bow wave in just as *Lively Lady* was taking the ground. She had lifted and settled with her keel too near the wall, thus laying outwards. She had no mast up and was secured to the quay by a breast rope from the outer chain plate. As the tide fell the load on the rope increased until suddenly it broke and the yacht crashed over.

However, *Lively Lady*, being the stout yacht she is, picked herself up on the flood when it returned.

At the best there was several weeks' work. All the interior joinery had to come out, the starboard water tank and all the carefully stowed stores had to be removed to get at the frames. Luckily the planking was undamaged, having given to the impact that cracked the frames. It would be well into October before she was ready to sail; far too late to attempt such a voyage. The part from England to Australia would have been quite all right, but it would have been midwinter in the Southern Ocean to round Cape Horn – not a pleasant prospect with the short days and long nights, the cold and the prospect of easterly winds.

There was only one thing to do, and that was to call it off for this year. It was a hard decision to make, but the only one. *Lively Lady* was hauled up into the shed.

Dorothy and David had to return to Portsmouth, and I saw them off at the station that afternoon. I stayed on to help clear the yacht of gear and stores. Mashford's put a store-room at my disposal, and we started clearing her out. It is surprising the amount of gear there is to move. I also stripped the mast of the standing rigging, and running gear, and rubbed the rigging down with boiled linseed oil before storing it.

Peter Lavers was put in charge of taking out the old, cracked frames and renewing them. Theo Mashford had a beautiful bit of oak hidden under a pile of offcuts, which had lain and seasoned for years. He said he had saved it for just such an occasion as this. The grain ran round in just the curve of the sawn timber, and it was as hard as iron. I wrote to the Lord Mayor of Portsmouth, telling him of the mishap and my decision to postpone my venture for one year. I asked him to let all my friends know through the Press, and this he did.

Before I had left Portsmouth an Irish actor friend of David's had given me a leprechaun as a mascot. He claimed he could undo all ropes that got snarled up round the rigging.

I got the idea that he was the cause of all the trouble. We got rid of him, but Dorothy will tell you in her chapter of how we did it.

On Friday, the 26th August, Francis Chichester arrived for his start on the 27th. He was lying alongside the jetty, taking on water, when he invited me on board for a drink. I was very impressed with his yacht, and the room in it. A bit too much width for rough weather; with no hand holds, one is apt to be thrown across the cabin. On the 27th he left, and I was one of those in a boat seeing him off. It made me feel very miserable to see him go, and me with *Lively Lady* crippled and in the shed.

I returned to my business in Southsea, but I paid several weekend visits to Plymouth to see how things were progressing, and stayed with Peter and his charming wife, Margaret. He was making a beautiful job of the repairs and *Lively Lady* was as strong as ever. I had the engine taken out and completely overhauled.

After Christmas I spent more time down there. I dismantled the winches and cleaned and oiled them. I rubbed down all the varnish work and revarnished it and repainted the interior of the doghouse.

Then came the job of re-stowing all the gear and stores. We got the mast out and redressed it with standing and running rigging. All this took many days' work. Finally we moved *Lively Lady* on to a trolley, gave her a final coat of anti-fouling and launched her. The mast was stepped and she was put out on to a mooring. The next week or so was spent in bending on sails, seizing and taping rigging-screws, etc. Finally, I bade farewell to my many friends at Cremyll, and sailed for Portsmouth.

The trip was uneventful, and I sailed into Portsmouth Harbour and picked up my moorings. It was the end of May, 1967.

6 *The Voyage Begins*

I SET myself the date of July 16th for the second start on my round-the-world voyage from Portsmouth. The last few weeks were hectic. The radio had been in the care of the makers all the winter and that had to be re-installed. Fresh provisions had to be obtained and put on board. A few days before the start the yacht was hauled out and another coat of anti-fouling put on. Water tanks were again topped up and fuel tanks filled. Several parties were arranged; one very pleasant evening was spent at the Royal Albert Yacht Club, followed by a small private dinner party at the Royal Naval Club. It was my birthday as well as Sir David Mackworth's, and we had a joint party.

I wanted a quiet start this year. I felt that this was a continuation of the voyage started last year and, as such, needed no great send-off. But my old club, the Eastney Cruising Association, insisted that I start from their club-house, and the Lord Mayor, Councillor D. Connors, made it known that he would be there in his official capacity to see me off.

The Royal Albert Yacht Club wanted to give me the starting gun from their Signal Station on Southsea front. I said I would be there at noon. When everyone is so kind what can one do but co-operate and help made a success of it? I had lots of nice telegrams and last-minute gifts of food and drink. I had willing helpers in my sister and brother-in-law – Muriel and Billy Goldsmith – and, at the last minute, Peter Lavers and his wife Margaret and daughter Julie rang up from Plymouth to say they were on their way. I had stayed with Peter and Margaret whilst fitting out the yacht during the winter. Peter had put in some good work at Mashford's repairing *Lively Lady*, and I was glad to see him.

I had to be ashore at Eastney by 10 A.M. on Sunday, so to be comfortable I arranged to sleep on board in Langstone Harbour. My friend, Mr Jack Spraggs, the Harbour Master, arranged a mooring for me. We dropped our mooring in Portsmouth Harbour on Saturday afternoon and sailed to Langstone. I was excited but, on this second attempt, I felt more confident. I considered the yacht was better prepared and I felt better in myself than last year. Algy joined at Langstone. Algy is a large, stuffed, white rabbit which David and Baba lent me as my mascot. He sailed the single-handed Transatlantic race with me, and has been inclined to be rather bigheaded ever since. He gives a lot of advice, some of which I ignore; but mostly he agrees with me.

Dorothy and I spent a peaceful night and, after a leisurely breakfast, dropped our mooring and went alongside a large yacht anchored off the Eastney C.A. club-house for that purpose. We were ferried ashore by the Coastguard launch, and met the Lord Mayor and officers of the club, who escorted me into the club-house. Toasts were drunk in champagne and short speeches made by the Lord Mayor and Commodore Rex Trye, to which I replied. It was a delightful occasion, happy and informal.

It was time to go if I was to reach the Royal Albert by noon. I invited Peter to accompany us and at the last moment Jane, my daughter, appeared together with her future husband Tony. They also came aboard. We cast off and headed out of Langstone Harbour, to the cheers of the members of the club and hoots of small craft. Some escorted us all the way back to Portsmouth. It was coming up to noon. Time to say goodbye to Dorothy, my daughter Jane, and my crew. Rex Trye came alongside with his boat and took them off. There was a crowd of small craft round, all calling out 'good luck' messages.

The yacht was under mainsail and light genoa and I let draw and headed for the starting line. The boom of the starting gun came as I crossed the line, and my voyage had begun. The club members gave me a splendid send-off and I had an escort

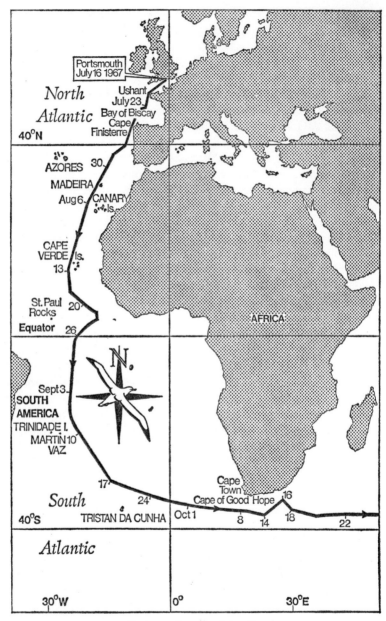

1 Portsmouth to Southern Ocean

as I headed for the Forts and the Nab Tower. The wind headed me so that I had to tack. The escorting craft dropped away and I waved a last farewell to Dorothy and Jane as Rex Trye turned back. I was alone, heading for Bembridge Ledge and the open Channel.

By four o'clock it was thick fog and I had lost sight of the Isle of Wight altogether. St Catherine's point foghorn boomed out of the stillness. Practically no wind.

I had some supper of cold ham, tomato, lettuce, and brown bread and butter. But it was not till then that I began to feel alone, utterly alone. As I realized what I had let myself in for I felt depressed and a little scared. Could I carry it through? The warmth and genuineness of all the good wishes came up before me. I mustn't let them down. Then there were those who really doubted my ability or who said 'What is the point of doing it now; it has already been done'; 'Your boat is too slow', etc., etc. I must prove them wrong. Besides, I had said I was going the year before, and only cruel luck had prevented it. I knew my boat was slower than *Gipsy Moth IV*, but so what? I knew I couldn't match her – boat for boat – but it remained to be seen.

The first night was dark and cloudy with a light southerly. I thought of my friends at home and wondered what I was doing, out here all alone, when I could have been in a nice warm bed. But with the dawn and the warm rays of the sun, life takes on a new aspect and things seem better. I had a cup of tea and some Shredded Wheat with hot milk. By this time it was flat calm and I was just drifting. Only eighty miles covered in twenty-four hours.

I had lunch of cold sausages, lettuce and tomatoes, brown bread and butter, and bananas. With the turn of the tide, at least I was drifting in the right direction. 'Ah, there you are – I told you she is too slow', I could hear certain people say; but who can sail without wind?

A faint breeze came up from the south-west, but with a short, lumpy sea, progress was still slow and with a head wind it was a succession of tacks. The second night came. I felt tired and weary, and turned in for a couple of hours. The last-minute rush and tear before the start of a long voyage is very tiring. Also I had a pain in my 'tummy'. I had had it for some time and I had fears of its being caused by an ulcer. I decided to eat carefully and to have a milky diet. By midday I had covered only another thirty miles, but during the afternoon a fresh breeze from the south-south-west enabled me to make better progress against the lumpy, short sea. Beating to windward in these conditions is not *Lively Lady*'s best point of sailing by a long way.

It was a dark night, with low cloud and rain. The wind veered to west-north-west, and I tacked, but feared I was heading too close to the rocks and strong tides at Ushant, the rugged island marking the extreme west of France. Then the wind backed to west and then south-west. I wrote in my log:

'What a place this English Channel is to get out of. I can quite understand the big square riggers getting towed down Channel.'

The night of Wednesday July 19th we were off Ushant in a flat calm, and the tide was sweeping us back the way we had come. We remained like this all next day and a bearing on the lighthouse put us as having drifted back ten miles. I reported to Niton Radio, and the next morning spoke to Dorothy on the radio telephone. It bucked me up no end, as I was feeling very depressed at our slow progress.

The sea was like a sheet of glass. It was sheer hard work trying to coax a little wind into the sails. Then a light easterly came, and I boomed out the light genoa to starboard. The wind was not enough to hold the sails still, but we were just moving in the right direction. A pigeon settled on board. He was obviously in a race, as he had a rubber ring on his leg, but he was very tired. I gave him a drink and put him under the cockpit seat aft. The wind varied from north-east to south-east and there was constant work entailed trimming sails to keep moving in the right direction.

On Friday night, July 21st/22nd, when we were about sixty miles south-west of Ushant, the wind suddenly went to south-west and increased. I hung on too long to the genoa, and it was a struggle getting it down in the dark. The hanks got jammed on to the twin forestays and I had to stand on the pulpit to free them. The wind increased to gale force, and we lay hove-to.

It eased a bit in the morning and we got going with the working jib and mainsail, but it was slow going against big seas. *Lively Lady* was throwing water everywhere. The wind increased again and I had to lower the jib and reef the main, but by the time I had finished I had to lower it altogether. Big seas with foaming white crests built up, and spray all over as I lashed the mainsail down. A small bird like a stormy petrel landed on deck. It seemed very exhausted. I put it in a corner under the spray hood, but I found it dead next morning.

It was a rough night, with heavy seas hitting us hard; but it eased in the morning and we got going again with working jib and mainsail. It was wet going though, with spray everywhere. Another pigeon landed. 'I shall be able to start a breeding pair', I wrote in my log, but what a mess they made!

By the afternoon it was near calm and I had the light genoa up. The sea remained lumpy, however, throwing us about and the wind out of the sails. The wind – what there was of it – went to north-east, affording a run. I lowered the mainsail and set the mizzen staysail. This sail will set when running, provided the wind is not dead aft; but the breeze was so paltry that not even this light sail would fill properly. We drifted more than sailed across the Bay of Biscay; the distance across from Ushant is about three hundred and seventy miles. It was the eighth day and my daily average was about fifty-five miles. One has to do big daily runs to counteract calms and head winds. I saw my hopes of a fast passage fading. The first of the pigeons took off. Having rested I suppose he felt fit to fly home.

I had the last of the bacon rashers, with fried eggs and fried potatoes, brown bread and butter, and a cup of tea. Bananas by now were black on the skin; but very good inside. The second pigeon became so tame that he would not stay out of the cabin and even landed on the dish of lettuce with tomatoes that I had for lunch. It was the ninth day and saw the last of the lettuce. I had enjoyed salad every day up till then.

At last I had a fair breeze and was getting along well under boomed-out genoa and mizzen staysail. There were several tunnymen about, with a forest of poles sticking out each side. They gave me a friendly wave. The D.F. bearing[1] on the morning

---

[1] D.F. bearings provide a method of fixing a ship's position by taking bearings of radio beacons from a radio receiver in the yacht. The radio beacon at Finisterre is a powerful one, with a nominal range of 100 miles.

of the July 26th (my tenth day out from Portsmouth), put me as about twenty miles north of Cape Finisterre, on the north-west coast of Spain – just about where I thought I was.

A big sea was building up astern and some, catching us on the quarter, caused us to yaw about a lot, with the sails getting aback at times and then coming back with a clap. The wind increased, white-crested seas with overcast sky and poor visibility. A large tanker, *Laplander*, passed on the same course as me. I hoisted 'M.I.K.', the code signal for 'Please report me to Lloyd's, London.' No notice was taken, so I tried to call him on the radio, but I could not make contact. By this time he was well ahead of me, and I had given up trying to contact him. Suddenly I noticed that he had turned and was returning to circle me. We made contact, and I asked him to report me to Lloyd's, London, and apologized for bringing him back.

By this time the wind had increased and, carrying the boomed-out genoa, we were overpressed. I got it down safely and set the working jib. Not without a scare, though, as when it was half up the wind got aback of it and I thought the booming-out pole was in danger. I hastily lowered it and tried again, with better results. I then found the top of the mizzenmast was bending alarmingly under the weight of the mizzen staysail, so I dropped that quickly. It fell into the sea, but I easily retrieved it. It would be silly to lose the mizzenmast and booming-out pole at this early stage; but it had been fun for the last hour running fast with the mizzen staysail up in rough conditions. We ran on under boomed-out working jib all night. It seems all or nothing with me. We were being hit by seas, some break-ing over into the cockpit. It eased in the morning, and I set the mainsail. Breakfast was Shredded Wheat and hot milk and boiled eggs with brown bread and butter. Then I changed the working jib for light genoa and we were slipping along at six knots. Lunch was corned beef, new potatoes, tomatoes, followed by very ripe and sweet bananas.

The night of Thursday, July 27th, was black, with the wind freshening and heavy black clouds to windward. We were touching seven knots at times, but I kept hanging on to the light genoa. At last I thought it ought to come down. With a strong crew it would be different, but I did not want trouble.

Lively Lady carried on comfortably under staysail and mainsail, until dawn, when with the easing of the wind I hoisted the genoa again. My noon position on the 29th put me as 39° 10′ N., 13° 16′ W, (about two hundred miles west of Portugal and four hundred miles north-east of Madeira), so I had to turn out the Azimuth Tables Book for latitude 39° to 0°.

I checked over the bread, and stripped all the wrapping off it. It seems to keep better exposed to the air. It was green on the outside but, after cutting this off, it was quite good inside. I had a pre-lunch drink of a can of beer, which went down very well, followed by sardines, cold potatoes, tomatoes, brown bread and butter, and a banana. This was the last of the bananas. They had lasted well – twelve days – and were perfect inside, although black on the skin.

We kept going well with a light, north-westerly making five and a half to six knots, and at dinner I had fried potatoes, tomato, and eggs, with bread, followed by a cup of Ovaltine made with milk. During the night I would make my special hot drink of a squeezed lemon and a large dessertspoonful of honey with hot water, topped up with whisky. This was a great pick-me-up in the dark hours – or at any time, come to that.

At noon on Saturday, July 29th, I notched up my first thousand miles in thirteen days. I had picked up a little on the average, but was still far short of the one hundred miles a day I had reckoned on.

The next seven days were frustrating in the extreme. Flat calms with the wind vane going round in circles; sails slamming on a painted ocean. We ghosted along, and I am amazed at the way we made some headway with no apparent wind at all. I wrote in my log:

'I'm resigned to my fate, the longest passage from Portsmouth to the Canaries.'

The Admiralty Pilot reads: 'N.E. winds blow steadily'. But not so for me.

During the night of Wednesday, August 2nd, Madeira Light was sighted bearing south-south-east about fifteen miles. I endeavoured to raise Madeira Radio – but no reply. By Friday the 4th Madeira was a hump away to the north-east. I tried again to make contact by radio. I could hear a ship talking and he picked up my call and reported me calling to Madeira, who then answered me. I asked him to report me to Lloyd's, London.

A light breeze from north-west sprang up but soon died away again, leaving us rolling about with sails slatting. However, the next day quite a fair breeze got up from the east and we made good headway, covering thirty miles in five hours, before it eased and we were doing five knots. I had the last of the tomatoes that day, hardly wasting any. The bread was nearly finished. I had to cut up a whole loaf to get some fingers fit to eat.

That evening I switched on the radio for a time signal and was startled to hear that a search had been on for me from Madeira, who had supposedly picked up an SOS from me. No trace had been found of me. I was furious. The message I had sent, merely asking to be reported, had been received quite plainly by them. No wonder they could not find me – I was a hundred miles away. I tried to make contact with Madeira, but failed to. Early next morning, however, I was successful and made contact to put matters right. I hoped they would broadcast it in the news so that I knew it had been received all right, and the folk at home relieved of anxiety. They did, and it came through later in the day that a signal had been received from me, saying I was not in any trouble. I heard later that another yachtsman named Rose had been west of the Azores and had sent out distress calls. Whether this was correct I have not been able to find out; if so, it was a strange coincidence that two people with the same name were sailing in the same part of the same ocean at the same time; but it is more likely that my first message had been misunderstood.

7 *Crossing the Equator*
For maps see page 46

I WAS settling down to sea routine by now, and it was almost a way of life. I had some good nights' sleep in my bunk, and it was much warmer – so much so that I practically stripped off to lie on my bunk at night; but I usually had to pull on my oilskin trousers to go on to the foredeck when changing head-sails. I always put on a webbed harness attached to which was a lifeline, although I must confess I did not always clip it on. When single-handed I often had to make several journeys between the foredeck and the cockpit to attend to the halyards and headsails forward and the sheet ropes aft. It is impossible to be clipped on all the time.

I got into a routine with meals, which were simple to prepare but varied and sustaining. I always fed well.

Our position on Sunday, August 6th, was 29° 05′ N., 20° 06′ W. At last we seemed to have picked up the trades, with the Canary Islands about 180 miles to the eastward. A light breeze blew steadily from the north-east and we were making about 5½ knots. I had a quiet night with a good sleep.

Quite a sea built up and *Lively Lady* rolled a good deal. As the day wore on the wind and sea increased, causing us to yaw about. At noon to noon we had covered 134 miles – the best day's run so far. We were tearing along touching seven knots when one of the steering ropes parted. I managed to knot it together and get her back on course. That earned me a can of beer.

The wind increased, and steep seas crept up astern, rolling us about miserably. That night I kept watch ready to shorten sail.

From midnight to 0300 hours we ran 22 miles. It was blowing force 6 to 7 and *Lively Lady* was burying her bows in a smother of foam. I had the two large genoas up boomed out either side for running. I thought it best to lower one, which I did. It was a struggle as all the hanks had unclipped themselves off the forestay and the sail billowed out into the sea. The noon-to-noon run that day was 150 miles, but it was rough going, and I was stiff and sore all over with the constant violent motion. The sky was overcast with low cloud. That afternoon I was considering renewing the steering rope of the Hasler self-steering gear which had broken, when I noticed the bumpkin which carried the pendulum gear of the self-steering was lifting with the pull of the servo-rudder aft. The tubular bumpkin is hinged on the aft deck, with a copper pin to hold it down outboard. This had sheered. With the speed of the yacht through the water there is considerable pressure on the servo-blade aft and upwards, and it was impossible to renew the pin without heaving-to. This I did. I found a brass bolt the right size and secured it with this. I renewed the steering rope and got going again.

I had a wonderful run during the next six days, covering some 950 miles, past the Canary Islands and Cape Verde Islands. It was hectic going at times, with big seas coming up aft, with foaming white crests. *Lively Lady* would lift her stern to them and sometimes they would break under her and she would surf along at what seemed fantastic speed.

The heat was terrific and I would be bathed in perspiration lying on my bunk. In the daytime I daren't go outside without my white floppy hat on.

I felt fit myself except for the slight pain in the lower part of my back – an old complaint that I didn't want to flare up. The pain at the start – the suspected stomach ulcer – seemed to have quietened down, as a result of my milky diet.

On August 12th in lat. 16° 27′ N. and long. 25° 34′ W. the trade winds failed – suddenly, just like that, nearly 1,000 miles north of the equator. It fell calm with a lumpy sea throwing the yacht about a lot. Then a breeze sprang up from the south-east. I stowed the boom on the genoa and sheeted it in, stowed the mizzen staysail and set the mainsail. We were close hauled, heading south. The next day I sighted the island of Brava in the

Cape Verde Islands, bearing south-east. Just a hump in the distance some thirty miles off. I had my first flying fish that day and it was delicious fried in butter.

The clipper route curved eastward now, following the curve of the West African coast, some 450 miles off Sierra Leone, before turning westward again to cross the line. This was the area of the Doldrums, and the wind went to south, so I tacked and steered east of south. Next it fell calm, the wind went to west and then north-west, what there was of it, and we drifted along more than sailed. I suppose this is typical of the Doldrums, but suddenly the wind shifted right round to south-east and blew up to force 6. I stowed the genoa and set the working jib and mainsail and we were close hauled again. I seemed to have found the south-east trade winds, and it was a sharp change from flapping sails and rattling gear to the whine of the wind in the rigging: but it did not persist.

On August 17th, 1967, my position was 9° 30′ N., 24° 0′ W. I had only made good about 300 miles in the last four days. As I have always said, one must have luck all the time to keep up these high average speeds, and how some do it I don't know. But at least I had kept moving, though at times we seemed to go faster than the wind. Although the wind was so light the sea, far from being smooth, was confused. On this day we seemed to be passing through an extra confused area and were apparently passing from the west-going north Equatorial to the east-going Guinea current.

The morning was dark and gloomy, not at all the kind of weather I had expected in the tropics. The sky was overcast with low black clouds affording hardly a sight of the sun. I had run into the south-westerlies and mind you they do blow – force 5, 6 or 7 in gusts. The sea was short and confused with breaking crests and *Lively Lady* thew a lot of water over her bows. The fore-hatch and ventilators had to be closed and this made it very hot below. The wind had a vicious whine in it and it was generally uncomfortable. I could just point east of south by compass – which with variation and a bit of leeway gave me about south-east. I wanted to get back to about 20° W. at about 5° N. before cutting across the south-east trades, to miss St Paul rocks and the South American coast.

August 19th was a day of mishaps. To begin with the deck
watch stopped. I had been too careful not to overwind it. Of
course I got time signals by radio all right, so no panic. But at
daybreak – after a rough night – on looking round on deck, I
was horrified to see the mainsail had parted from the slides on
the mast. In the half light I thought at first that the slides had
been pulled out of the track, which really worried me. However,
they were there but the seizings had worn through so that the
sail had been torn away from the slides. I lowered the mainsail
and set the trysail while I re-seized the slides. This took me a
long while – and at one time I had to give up because of being
thrown about the deck so much by the seas. *Lively Lady* rides
quite well with the trysail and in any case I could not have
carried the main for long in these conditions. Then I had some
boiled potatoes, onions, and chicken suprême turned out in a
deep bowl for lunch. I turned my back for two seconds, when
lurch and the whole meal crashed down and was smashed. Oh
well!!

I had not had a proper sight of the sun for several days and on
this day I got only one hazy shot which gave me one position
line. My dead reckoning had been pretty good to date – but
bashing to windward in a short sea at three or four knots with
an east-going current as well is liable to give you funny results.
I was very fed up and depressed at the slow progress, and with
the rough handling the yacht was handing out to me.

Today I had to throw my first egg away. I had eaten three or
four a day and now had only about thirty left. I thought it
best to eat them first. Potatoes and onions were good, and so
were the small whole cheeses. As I have said, I had an apple
after every meal and a lemon a day and sometimes a grapefruit and orange as well. They go down well in the heat and
provide the necessary vitamins.

I quote now from my log and diary written at the time:

August 23rd. Have continued our way south-east in much the
same conditions but this morning with the sea down and a little
sun conditions are better. Picked two flying fish off the deck for
breakfast. At 4° N. and 18° W. I decided to go on port tack and
head west-south-west. This should take me over the Equator (we
hope) at about 25° W. and well clear to the east of St Paul rocks

and no fear of getting too near the north coast of South America. Feeling in better spirits as we begin to show progress. Ships have been known unable to weather Cape St Roque, which is nearly the most easterly point in South America.

August 26th. Well, we eventually crossed the Equator during the night in long. 23° W., which was better than I expected.

This was a great occasion for both Algy and me, for neither of us had crossed the line before. We toasted Father Neptune in a hot whisky and invited him on board to share it. Algy was a bit scared in case he was ducked in the sea, having an aversion to getting his feet wet. However, being night and much cooler I let him off.

No other celebrations were made. After all, being single-handed one day is much the same as the next. Great fun is had in big ships with crew and passengers joining in the ceremony as they did in the old sailing ship days. In spite of being on the Equator it was much cooler and the nights were longer.

August 28th. The wind gradually went round from south-west to south, but not to the steady south-easterly which it should have been in this region of the trade winds. We are heading south-south-west, but what with leeway and a westerly current of up to two-and-a-half to three knots we get carried to the westward. That is why it is so important not to cross the line too far west. This makes several weeks of close-hauled plug and I must confess I didn't realize it would be such hard work. With any wind the sea gets up in short lumps and we bang and crash into it – throwing water everywhere, which always slows us up. I have made one or two attempts to contact a ship but I am right off the 'lanes' and the nearest one could be several hundred miles away.

Food is good and plenty of water in hand. Today I had to throw three eggs away. Until now only two have been bad, though one or two have tasted a little stale. I have a dozen left so I have done very well with them. After this I suppose it will be dried eggs. The apples, oranges, grapefruit, and lemons are still good. Very few flying fish have come aboard and only on two mornings could I have flying fish for breakfast. I have seen very little wild life. Not like the North Atlantic. I am a little disappointed at our speed across these trades – but it is the seas that stop us. I could go faster, I suppose, by sailing a bit more off the wind, but as I said the sea is pushing us as far west as I need. It would be different if I wanted to reach a South American port.

August 29th. Felt seasick this morning for the first time since I left. Shows what motion we go through. What a rough 24 hours it has been, and people ask, 'What do you do with yourself?' Twice I had to turn out in the dark during the night to reef the mainsail and then to take in the genoa staysail. This means struggling into oilskins and safety harness, an awkward job with the violent motion of the yacht. I must confess that I did not always get my safety harness on before going on deck. There was not always time.

All today we have been under reefed mainsail and working jib – making about three to four knots against a south-south-east with really big seas. The motion is sharp and jerky and catches one un-awares and throws one right off balance. Every smallest job therefore is a thought-out plan of action – taking four times as long as normal. This goes on all the time – every minute – and is rather wearing and tiring, coupled with the constant whine of the wind in the rigging. One had only to have one or two days like this to knock the daily average right back, although I would have been well on if it wasn't for the calms of the first three weeks. It was then that all chance of good time to Melbourne went by the board.

August 30th. Our position is 6° S., 24° W. and we are about 600 miles east of Cape St Roque. Hauled the log line in. For some time I have reckoned it was showing slow. Just as I thought – a whole row of barnacles all along the line – it must have slowed it down a bit. The barnacles are the great long ones rather like a bean on a stalk.

Have discovered it's no use trying to force the yacht along too fast against the sea. We only get a bashing and stop and then start all over again. The lee rail under in smooth water may be all right but not under these conditions – so I have left the mainsail with a couple of rolls in it. We go just as fast and not so much violent motion, water on deck, or wear and tear on the gear. Keep getting a call sign of 'ASN' on the radio, so assume it's a radio beacon on Ascension Islands which are mid South Atlantic, half-way between Brazil and the Congo. Can't pick up a ship yet to let people know I'm all right. This is the thing about this trip – am I being too selfish in giving my loved ones these long periods of anxiety?

Cooked the last three eggs for breakfast and all were bad but I don't suppose I have thrown away more than a dozen, very good really.

September 1st. It beat me in the end – it always does. I was forced to lower the working jib and set the storm jib. The seas got bigger and more confused and we were standing on our ends. The yacht and gear were taking terrific punishment – to say nothing of me. Black storm clouds loomed up and the wind gusted up to force 8. So this morning I lowered the spitfire staysail. We lost a few miles but it's no use carrying something away out here as I'm a few thousand miles from home now. Later in the day I got the spitfire up again. Now at least we make some headway in the right direction. I was hoping to pick up a little time down here. (As I write a big sea hits us – spills over and fills the cockpit.) More low black clouds drift by – nearly touching the masthead. Altogether it's wild and wintry looking, though we are not far south of the Equator.

The wind was south-east and we were heading west of south. The chart showed a westerly current so we were probably making south-south-west. To crown it all I got a touch of lumbago.

On September 2nd after a wicked rough night we were under storm jib and reefed mainsail heading into the sea which was hitting us hard and giving us a very rough time. What a bashing we had had during the last two weeks. It seemed never to cease and I was stiff and bruised. Daylight showed great banks of black cloud out to windward. The shriek of the wind was constant. I had a mood of depression this morning and my nerves were so strung up that I found I could not make myself do any jobs. One had to be a trapeze artist to hang on.

It was impossible in the rough conditions to take sun sights, but my dead reckoning put me at 10° 30′ S. and 25° 0′ W.

The clipper route keeps west, to avoid the area of variables, to the north-west of the Cape of Good Hope. Then it gradually turns eastwards to pass just north of Tristan da Cunha in mid South Atlantic, and past the meridian of Greenwich – long. 0° on lat. 40° S.

My log says:

September 9th. This last week we have made quite good headway south. We left Trinidade Island about 300 miles to the westward and some 900 miles off the coast of South America, which was more than I had intended. However, southing is the important thing. We had a couple of good days but then last night a flat calm

to spoil it all. This is an empty barren ocean and I haven't seen a
sign of life of any description until today I saw what I presume is a
Cape pigeon. I haven't seen a single shark's fin all the way and
only a dozen or so dolphins. One could be out here for ever and
never be seen. I've had two days of sun – which is wonderful for
me. At 22° S. I shall be running into the variables soon, as we
curve away south-east (or try to). As I get towards the latitude of
South Africa, I am hoping I may pick up a ship to let people know
I'm all right. As I have said, it is this that is worrying me.

At dawn on September 16th it was blowing hard from the
north-west and we were running under storm jib and trysail.
During the morning it eased and I changed the storm jib for the
boomed-out working jib. I was fooled for at about noon a sudden
squall blew up and broke the booming-out pole. It just buckled
up in the middle. *Lively Lady* came round to windward. I
dropped the jib and got her back on course, running under
trysail. I hoisted the storm jib. Big white-crested seas hit us
hard. By nightfall it was blowing force 9, and I lowered the
trysail as she was overpowered. 'What a terrible night', I wrote
in my log. At about 10 P.M. I had to take the tiller as the self-
steering gear wouldn't hold her. I assumed the vane was
slipping. During a lull I tried to get it going, but was horrified to
find that the servo-blade had snapped off. *See figure, p. 183.*

Conditions became worse with winds up to force 10 and
heavy seas swinging the stern. I lowered the storm jib and lay
a-hull and then retired below to a hot drink and a rest. Dawn
revealed a terrible sight, with great white-capped seas and
wind still up to force 10. I didn't relish the job of repairing the
vane gear, but I got the servo assembly inboard and fitted a new
servo-blade. The vane had to come off – but it was impossible
to repair in those conditions. During the afternoon I made
several attempts to go aft but the wind was such that one
literally had to hang on. Big seas hit us and one in particular
broke right over us, laying us well down. I gave it up for the
night. Brilliant lightning lit up the scene.

I had a hot drink of lemon, honey, hot water, and whisky. It
blew hard all night but at dawn it eased. I then tackled the job
of unshipping the vane. It was exhausting working on the stern
as it heaved to every sea. I made sure my lifeline was securely

clipped to the mizzen rigging so that I could not be swept overboard. The stainless steel tube on which the vane is clamped was slipping in the clamp at the bottom. I took it out and roughed it up and re-clamped it. It was fortunate that I had taken a very full kit of tools with me; without them it would have been impossible to have put right the self-steering gear and carried out other repairs at sea. The breast drill was particularly useful and I could not have done without the vice.

By midday we were nearly becalmed but making some headway in a light south-west breeze, under all working sail. On September 23rd the sun crossed the Equator to 'S', so it was officially spring in the Southern Ocean. My position was 34° 45′ S., 6° 15′ W. I had been heading south-east but there was a strong current setting against us north all the time when I wanted to get more to the south. In two days we were set thirty miles further north.

I noticed the top slides of the mainsail had come adrift. I lowered it, the slides remaining up aloft. I changed the mainsail for the old one, which was slightly smaller, and, I thought, better for these conditions. It seemed to set better too. Changing a mainsail is not an easy job on a pitching deck and then bagging it up to get below and stowing it away. Everythink on deck was coated with salt, which brushed off on to my clothes in a white powder, a nuisance because it made them damp.

The wind went to south-east and it was tack and tack to gain a few miles against lumpy seas, until September 27th when it went to south-west force 3. By nightfall, however, it had increased to force 7 and I reefed the mainsail and stowed the staysail. Conditions got worse, and by midnight it was blowing a strong gale, force 9, and I lowered everything and put up a small stormsail. This just enabled her to answer the helm and stopped the pounding under the stern. The wind increased to storm force 10, and dawn showed big seas with white spume down their face. The wind roared through the rigging, and the whole ship shuddered and quivered as big seas struck her hard and broke over, sending clouds of spray everywhere.

I went forward and lowered the stormsail in a hailstorm which tore at my face and blinded me. The hail hit the deck like machine-gun bullets. This lasted all day. We were tossed and twisted about like a cork and all we could do was to take it, and drift north-east, miles off our course. A second terrible night followed and once we were hit so hard I thought that we were stove-in and went round with a torch expecting to see water coming in. But all was well.

By dawn on September 29th it eased to force 8 and I managed to set the spitfire staysail. The wind backed to south-south-west and we just managed to head south of east. By midday it had moderated to force 7 and I set the trysail. Big seas were still running, knocking us back, and after 36 hours of this fierce gale I wondered where we were. We must have drifted considerably.

The next week the weather continued to be miserable, with heavy cloud and misty rain, and winds varying between flat calm and force 7 or 8. It never seemed to blow steady; from calm it would go to strong in no time. With the big seas and cross seas it was like riding a bucking bronco, only our ride was for 24 hours a day. During one blow I went aft to adjust the self-steering, when I noticed some set-screws were missing from the servo-box and others were loose. The whole thing was in danger of collapsing. I hove-to, got the servo assembly inboard, not without a struggle, and repaired it. Luckily I found some small bolts with the same thread to replace the lost set-screws.

While I was doing this the wind increased to storm force 10, and I had to lower everything hurriedly. It was the fourth full gale in a week. Big seas swept us, tons of water coming over and filling the cockpit. The motion was severe and to crown it all I had an attack of lumbago. This is awkward, to say the least of it, in a small yacht, in a gale and alone. As any cruising man knows, if he has suffered from this disability, lumbago makes it difficult even to get out of a bunk. Any sudden movement causes sharp pain, and at sea movement cannot be avoided. Lumbago is a minor complaint ashore but at sea it can be crippling.

8 *Gales in the Southern Ocean*
For maps see pages 46 and 64

FRIDAY October 6th was a great day when I managed to contact Cape Town radio and send a message through. This made me feel very much better. The fact that it was blowing a force 9 gale didn't seem to matter.

My position then by dead reckoning was 37° o′ S., 14° o′ E. which put me about 300 miles south-west of Cape Town. Before I left England a reporter from a South African newspaper had telephoned me, inquiring whether I intended calling at Cape Town as great interest was shown there in my plans. The radio operator was also very helpful and assured me of a welcome in South Africa, but this was impossible because time was precious if I was to get to Australia and from there on round Cape Horn before the southern winter set in.

The next day I charged batteries in order to be sure of power to speak to them again. But horror struck when I went below to find the cabin full of smoke and the switchboard burning away. Panic stations then to switch everything off. I found a short behind the switchboard where the wires were connected. This of course had *dis*charged the batteries. However, I managed to patch it up by moving the wires to a new position on the board, and get some more charge into it, but I couldn't spare current to use my lights. After several fruitless attempts I got Cape Town again on the 9th and received a nice message from Dorothy. This bucked me up no end. Even though it was a day of flat calm with not a ripple. I did odd jobs and decided to fix a new booming-out pole in place of the one that collapsed. This meant going nearly halfway up the mast – and on

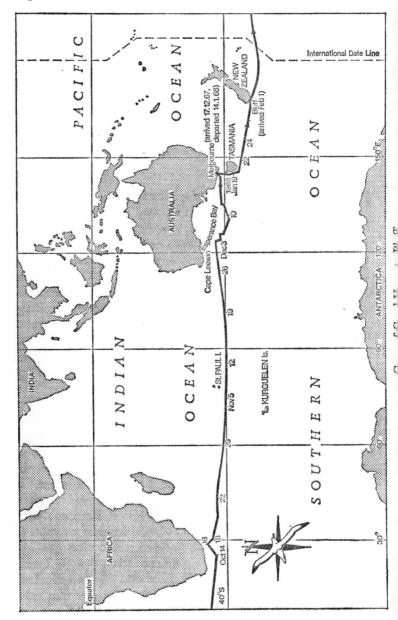

one's own – taking tools, etc., getting the old one off, coming down and going up again with the new one. I found I could just reach by standing on the boom and then on the headboard of the mainsail raised a little by the halyard for the purpose. It meant undoing a nut with a split pin through it to take the old one off the slide and putting the nut and split pin back again. Not an easy job, clinging round the mast with one arm and one leg.

I must have strained myself as that night I was in agony in my back and lower stomach and particularly in the right groin. I took some tablets and lay in my bunk actually sweating with pain – not daring to move a muscle. The next day all I could do was to heave myself out for a drink and go back to my bunk again. I really began to think I would have to alter course to Cape Town – if I could! I wondered if I had ruptured myself or something like that.

However, the next day I was very much better and the pain gradually cleared away.

It was then that I got Cape Town radio and a message from my daughter Jane to say she had married on October 7th; bless her. On Friday the 13th it blew a gale from the east and from then on it was gale after gale with calms in between and me fighting to get east, when we should have been getting the prevailing westerlies. One day I felt very exhausted and weak and every job was an effort. One very severe south-east gale pushed me much too far north and then when I got another strong gale from the east all I could do was to head her south.

I now lost contact with Cape Town but picked up Port Elizabeth, some 400 miles to the eastward, though they were not so strong. I had messages from Dorothy and friends at home.

At last on October 17th it was blowing a strong gale from the east and gradually backed to north-east. This allowed me to head east-south-east and so get more like a course I wanted: to get east and also farther south. It was a wicked day, though, and one albatross – swooping around as they do over the sea without effort – got carried on to the forestay and fell into the sea, just a bundle of feathers. I hadn't time to see what happened to him. I was too busy getting the jib in, with solid sheets of water drenching me – running into my eyes and my mouth.

The flaying halyard caught me one in the eye and blacked it! The deck had been running with water for weeks and it was impossible to go out of the cabin even into the cockpit without oilskins on.

The wind stayed north-east and we made good progress until October 25th when it was blowing force 8 to 9 again and I was obliged to lower sail. Nevertheless during the last week we logged over 900 miles, which was encouraging. If it hadn't been for those long continuous calms in the early part of the voyage and then round the Cape of Good Hope, I should have been well on the way by now.

It was miserable down there with thick fog – rain – gales – heavy seas – no birds – no nothing. A hundred days at sea found me reasonably fit, and morale was better now. Algy chattered away, always answering me back. He slept in the other bunk, in fact he stayed in it the whole time. However, for a week or so at this time he had been continually grumbling. Everything was wrong and at last he came out with it. He demanded that we alter course for Cape Town. He said he wanted to see a bit of night life. In the end he got so agitated that he fell out of his bunk and I had to clap him in irons in the fo'c'sle to cool him off. It was sheer mutiny, and it wasn't until I threatened to hang him from the cross-trees that he quietened down.

I was still eating fresh fruit, which was good. That and the vitamin pills from Doctor Eddings and Mr Tremlett, a local chemist, plus a little alcohol and a varied diet kept me fit. There was an empty Courage beer can nearly every day along the route – so I could easily be tracked.

The onions had long green shoots on them by now, and I ate them with biscuits and cheese, keeping the natural Vitamin C going.

I still found it difficult to get south. There must have been a northerly set to the current, as although I was steering south of east I found myself further north. It was rough going, but we were getting along well. On the 21st I had spoken to Port Elizabeth Radio at about 650 miles, the last contact I made with Africa.

October 25th found me fighting a north-easterly gale, force

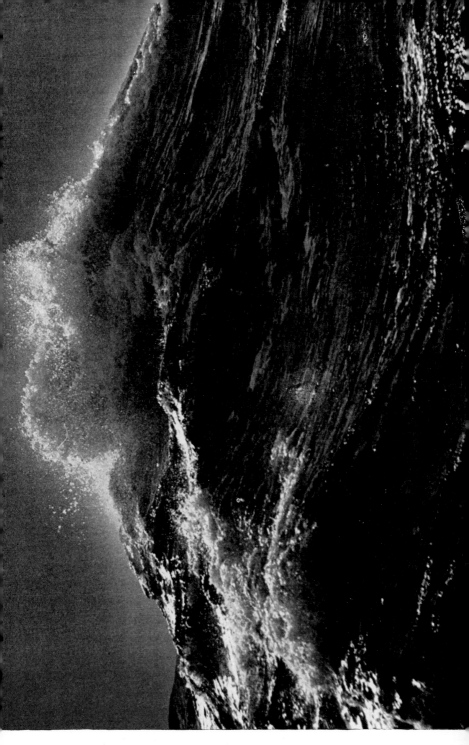

13. A glint of sunshine lights up a magnificent ocean wave as it breaks.

14. An Australian escort for *Lively Lady* off Williamstown.

15　Greetings off Port Phillip Heads on the approach to Melbourne.

16. On arrival in Australia, Michael, the author's second son, with his wife Judy, and their two sons, Christopher and Nigel.

17. The author working aloft in a
bo'sun's chair.

18. Approaching New Zealand for
masthead repairs.

19. (Above) *Lively Lady* enters Bluff harbour in a gale, and (left) 20. her skipper looks tired after a difficult landfall, following damage in frightful weather.

21. The sea boils as the tops of the waves are blown off.

22. Cape Horn stands up steep and grim above the heights of Tierra del Fuego as *Lively Lady* crept past on 1st April 1968.

8 to 9, obliging me to lower the reefed mainsail. It was hectic for a few minutes when the shackle connecting the main halyard to the mainsail got caught in the jib halyard up aloft, and I couldn't get the mainsail down. It flogged about for some minutes, before I could free it, breaking a batten and tearing two mast slides off. We carried on under spitfire staysail. We had had these easterlies for two weeks now, and I began to wonder where those westerlies, that are supposed to blow constantly round the roaring forties, had got to.

The next day it eased enough for me to replace the broken batten and to re-seize the slides to the mainsail.

It was bitterly cold and I had three sweaters on. The wind backed to the north-west on October 27th in the early hours of the morning and I had to dress in oilskins to go aft and adjust the self-steering gear. We were running under trysail and spitfire staysail with the wind increasing. Dawn showed a wintry scene with a wild boisterous sea causing us to roll heavily. Low, heavy clouds swept by. By midday it was gale force 8, but we managed to hold on to the trysail and spitfire staysail, running at from five to six knots, in the right direction. The glass was still falling and during the afternoon hail and sleet squalls descended on us, laying us well down. Dirty-looking low black clouds blew up from windward.

What a cruel night that was. The darkness could be felt. A flash of white would show up as a white-crested wave ranged up alongside. Hail, sleet, rain, and cold – the worst night I had experienced. As usual, the sea won, and I was forced to lower the trysail in a fierce hailstorm biting at the skin on my face and wind force 12 shrieking and tearing at me in the pitch blackness. The same thing happened an hour later when I was forced to lower the staysail; the sail flogging and shaking the mast and the ship. I thanked God for the dawn, but that only showed up the wild wintry scene – visibility down to 50 yards and the deck white with hail and sleet.

I wondered if I had carried on too long with sail. What is the borderline between being bold and daring and being foolish and silly? This was a desolate part of the ocean, and it would have been silly to have carried something important away at that stage.

R—F

I retired below to a hot drink. As everything was damp I lit the Tilley heater, and it was warm and comfortable. All hell was let loose outside, and it continued all that day until next morning, when it eased and I got under way with reefed main-sail, spitfire staysail, and working jib.

It was still squally and there was a lumpy sea, which at one time threw me across the cockpit on to a sheet winch, bruising my ribs. I had to watch constantly as the wind would die away to force 2 and then increase to force 6 or 7 suddenly, sending me scrambling to the sheets.

I found the cross-link between the latch gear and the servo tiller, carrying the ball-joint, broken off that morning. This was serious and I wondered how I could repair it. I got the gear inboard and took the broken pieces off. It was a hollow tube. I found a piece of brass rod that just went in with a driving fit. I cut it down and joined the two ends together, and it was stronger than before. *See figure, p. 183.*

The self-steering gear is set for a certain wind strength to hold the yacht on course and with a sudden great change in wind strength it cannot do this. That night I had to scramble out constantly to hold the tiller myself in the blackness that could be felt and the deck running with water. One has prac-tically to live in oilskins as it is impossible to go into the cockpit to empty the gash bucket without them.

November 1st was a sunny day with wind down to force 1 or 2, but a big and irregular sea was running and I had the genoa up – boomed out – and the mizzen staysail – but the sea rolled us over and threw the wind out of the sails. Nothing is ever right is it? I hoped for a few days' fine weather with a fair breeze to catch up a bit of time. There was still about 4,000 miles to go, and when I thought of that and the fact that it was 2,000 miles back to Africa, I was careful about the gear. I was on my own – with no radio contact – and to carry away sails or gear would be serious. So although I sailed the yacht constantly at the best we could do in the conditions, and occasionally carried on too long, I didn't think it wise – good seamanship – or fair on my own strength to take undue risks. To arrive properly and in condition was the thing I aimed at.

It was gale after gale and the night of November 5th saw the wind get up to force 7 then force 8. We were running under boomed-out jib and I hung on to it too long. The yacht was almost overwhelmed and I had a hectic time getting the sail down and disconnected from the boom. The roar of the wind, the whine in the rigging, the hiss of the sea, the spray coming over, made so much noise that no one could hear me singing 'Onward Christian Soldiers'. I couldn't even hear it myself, and I could see nothing in the pitch blackness.

At dawn I was surveying the forlorn scene of the endless succession of waves, when I was startled by the sight of a huge whale surfaced right alongside me. He was vast and had a mottled grey look of age about him. He 'blew' and I got the scent of him. He lay in the trough of the big seas and I thought of his great strength and power – symbolic of the wild, cruel Southern Ocean. Only he could meet it on equal terms, I thought as he dived and disappeared.

Later in the morning the sun came out. What a difference that makes! I set the boomed-out storm jib on one side and boomed-out working jib on the other and with a west-north-west wind we were making good progress at six knots. White-capped seas followed us up astern. It was a quiet night and *Lively Lady* ran on while I had a good sleep. I had my last apple. The last few had started going soft, but I didn't throw away many. Oranges, lemons, and grapefruit were still good, also the onions. I ate the green shoots with biscuits and cheese followed by honey, butter, and biscuits and a good slice of cake for supper.

9 *Nearly Dismasted*
For maps see pages 64 and 76. Sail Plan page 176.

THE morning of November 7th was fine with wind and sea moderate. I went forward to change the boomed-out jib for the big genoa. I stood at the foot of the mast with one hand on it, when it gave away from me and came back with a clang. At the same moment the lower backstay fell down round my shoulders. I immediately let go the jib halyards, and let them drop. The mast, which is stepped on deck, was banging in its socket. Only the topmast shrouds were holding it from falling over to port. It was whipping in the middle and coming back with a clang. I looked round for something to secure it and quickly unshackled one of the staysail halyards and let it aft to an eye-bolt and winched it up. Then I did the same thing with the other one and also with one of the twin topping lifts.

I surveyed the damage. The lower backstay had broken at the splice at the top end. Some strands had pulled and it had drawn round the thimble, leaving that aloft. Then one of the staysail halyards, which acted as a jury backstay, went with a bang and the topping lift pulled the sheave off the mast.

I unshackled one of the twin inner forestays and shackled that down to an eye bolt with a rigging screw to tighten it up. I also led one of the jib halyards aft, under the cross-trees, and secured it. Talk about panic stations! I shall never know how I managed to save the mast.

Then I got the backstay aft and began to repair the end. I cut off the frayed part and bent it round another thimble, securing it with bulldog grips; but, while I was doing this, it began to blow up again. The yacht was lying with the weather on the starboard side. This was the side with the broken back-stay, and I tried to turn her round so that the weather was on the other side and the weak side to leeward.

In the end I started the engine up and turned her round with that. By now it was blowing a gale, with a fierce hailstorm. I finished repairing the end of the stay, wondering how I was to get it back up the mast. It was hopeless trying to do it under those conditions, so I settled down for the night. It was a wretched night: pitch black, blowing a gale, raining and cold. I sat or lay on my bunk, tense, thinking any creak or bang as the sea hit us meant that the mast had gone. I lay all next day like that, until evening, when it quietened down a bit.

I rigged a tackle and hoisted it up the mast with the main halyard. I put the bo'sun's chair on the end and attempted to hoist myself up; but the motion was too violent, and I was swung about and bashed against the mast. The tackle wasn't right either. I had not enough purchase. I retired below, very de-jected, and made myself a cup of tea. I prayed for a calm day tomorrow.

Sure enough, it was. But of course there is always a big swell, which catches *Lively Lady* now and again and rolls her. This time I hoisted the mainsheet to the truck on to the main halyard and with the aid of this purchase, pulled myself up with that. I arrived up there: panting, with my thighs sore and the grip in my hands weakening; but I did it, and secured the stay and, with shaking fingers, split-pinned the cotter pin.

Back on deck I went on my knees and thanked God for helping me and giving me the strength to do it. I tidied up on deck and we finished up minus one staysail halyard and one topping lift. It was a very lucky escape. I retired below to hot coffee and rum. The porridge was on and I felt content. After a quick breakfast we got under way with boomed-out jib and reefed mainsail – cautiously at first, I must admit.

We were just about halfway between South Africa and Australia, with three thousand miles in either direction. It would be disastrous if I got disabled there. Just to the north of us lay St Paul Island: a small, rocky island with a few goats on it, I believe. The chart states that a food store is kept there, with clothing for shipwrecked mariners. I had visions of making there and waiting for a ship to call, which could be months.

On November 12th I sighted Australia – that is, on the chart. The chart of the Southern Ocean is just a large sheet of white paper with South Africa on the left and Australia on the right. In between there is nothing except latitude and longitude lines and a row of X's marking my course. It was folded down the middle to fit my chart table, and I had just had to turn it over, so I was really making for Australia at last.

Day followed day with the same pattern of endless white-capped hills of water – all intent on our destruction. The moan, the whine, the whistle of the wind still there. The clatter of spray on deck and every now and again a crash as a heavy one hit us solid, shaking even the mast. But we made headway, covering 800 miles in seven days.

On November 15th we were running under twins boomed-out, when the wind increased and I went to lower the genoa. I found the hanks had engaged themselves on to the other forestay and I had to lower the other sail as well to unhank the genoa. On getting it down, the slide on the mast broke away at the welded joint and the boom came away. This meant I had only one boom which could only be used one side.

I had another attack of bad back, so painful that I had to shorten sail and lay in my bunk with some Codis tablets. I was in such pain that I was sweating and hardly able to move. Nothing exhausts one so quickly as pain. This lasted two or three days and I was thankful when it cleared up and I felt better. Of course the usual severe gales occurred. On the 17th, on looking round in the morning, I found the bobstay adrift from the eye-bolt fitting on the bows just below the waterline. I rigged ropes and bo'sun's chair to go over the bows to try and repair it. I got down there, but decided it was too danger-ous. I was being dipped every time *Lively Lady* dipped her bows and I couldn't reach the eye-bolt. It was a struggle getting back on board, so I left it. It made me realize what an almost impossible task it would be to climb back on board from out of the water if I fell overboard. The bowsprit is very short and strong and I rigged a wooden strut between the bow-sprit and pulpit. Just extra support.

It was at this stage of the voyage that weaknesses begin to show up from wear and tear, but we could soon put them right when we got to Melbourne.

A small Japanese motor-vessel circled me, but before I could get his name, or signal him, he cleared off.

November 21st was another dreadful day.

At about 2.30 a.m. I was roused from a sleep by a banging noise. I rushed on deck to find the other lower backstay had gone in the same way as the first. The mast was doing all sorts of 'S' bends and coming back with a clang. I only had the working jib up on a boom. It was the same frantic pattern of moving a forestay (the upper one on this occasion) and one of the jib halyards round under the cross-trees and securing the mast, only this time it was in the dark and with a freshening wind.

But I saved the mast again. How I did it I shall never know. Before I had finished the wind was blowing gale force 8 with seas and spray coming over. I set the spitfire staysail and watched with bated breath. The seas were bad and laying us down to leeward, so I thought it would be no worse running with it. I was wet and cold and dejected. Only the previous day I was beginning to work out an arrival date. Now it could be any time. It was impossible to set the mainsail, as the forestay still led under the cross-trees and across the track up the mast. We ran like this for the rest of the day and until next morning.

The wind then dropped to light westerly, but the sea was still high and irregular. It would have been impossible to try to hoist myself up the mast in those conditions. I hadn't yet tried to repair the broken end of the stay, as I didn't even know whether I had enough bulldog grips.

On looking round that morning I found that the servo-shaft of the self-steering gear was rocking about in its box, and on inspection found that a bronze pin on which it swivels had sheered. The damage looked serious, as at first I didn't think I could repair it. However, I got the servo assembly off and found a bolt of the right diameter. I cut it down to size and screwed it in, with a locking nut on it. The result made it as good as ever.

It was a sunny day and it was good to feel the warmth of the sun. I set the boomed-out working jib and the mizzen staysail. But it felt flat calm and I lowered the mizzen. I found some more bulldog grips and repaired the end of the stay. The sea was still very lumpy, however, and we were rolled about all over the place in the swell. Impossible to attempt to go up the mast. I felt at a low ebb, morally, and unable to attempt it.

The next day, the 23rd, was grey and overcast with fine rain. I moved one of the inner forestays round the mast and shackled it to the main topmast shroud plate, for extra support. With a fair breeze from the north-west I set the mizzen staysail and we got along steadily at five knots. We ran like this until next morning when it fell calm, and I lowered the mizzen staysail. The sea also was reasonably smooth, and it began to dawn on me that now was the time to have a go, and put the backstay up.

I consulted Algy and he agreed. There were several alba-trosses about and I could hear the voices of old-time sailors, who sailed these waters, taunting me and goading me on.

'Call yourself a sailor,' one of them said, 'we would have gone up.'

'It's all right for you to say that,' I answered, 'you had a big ship and lots of you.'

'We still had to do it, though,' they said,

'Another thing,' I replied, 'you had nice firm ratlines to put your feet in, and a ship that didn't roll with every ripple.'

At that they were silent, but they had put me on my mettle. I rigged the mainsheet on the main halyard as before and hauled myself up, and pulled the backstay up after me. I also pulled up the topping lift and lashed a block to the cross-trees to take it.

Back on deck I was very pleased with myself. I cleared up, put the forestays back forward in their proper position, and tightened up the backstays. We were shipshape again, but with two suspect lower backstays that would need watching. We could at least set the mainsail. By the time I had finished it was late afternoon. I had had nothing to eat, so cooked myself a hot meal and with a hot drink I felt better. I left the yacht under jib for the night and the next morning the wind shifted to north-east. I set the reefed mainsail. The wind increased to force 6 and I inspected the lower ends (I had reversed the ends) of the backstays for any sign of the bulldog grips slipping, but they appeared to be holding all right.

I was startled to see a ship approaching from astern. It was the first I had seen since one passed me by the Canary Islands, other than the small Japanese motor-vessel. I hoisted M.I.K. signal, but she passed with no sign of life on board, no answering signal or toot on her siren. Perhaps nobody was on watch.

By now, November 27th, I was on the same longitude as Cape Leewin, the notorious cape at the south-west corner of Australia – notorious for its gales and high seas. I was some 300 miles south, however. It was squally and one hit me suddenly, knocking *Lively Lady* almost flat, and sent me slithering to the halyards to let go and claw the mainsail down, and then the jib. The sea was a mass of white foam, and breaking white across the deck.

I left her under spitfire staysail. We were being rolled down unmercifully and I eyed the suspect backstays. It was frightening to look out over the sea, as great waves rolled up, with their tops being blown off in clouds of spray and mist. The wind increased to storm force and I was forced to lower the spitfire staysail in a fierce hailstorm.

I was farther north than I had planned but after the backstay trouble I edged more north towards South-west Australia in case I had to make for Albany for repairs, although I was hopeful again of reaching Melbourne. But I don't know of anything so frightening and sickening as to see the mast doing figure-of-eight bends. Quick action was necessary and I was rather pleased to see I had still a mast standing. If the mast went I would have done all that one man could do to prevent it, I felt sure.

As the storm subsided I got the spitfire staysail up again, only to find the servo-blade of the self-steering gear pushed back and splintered at the top and the brass pin sheered. That had to come off, be got inboard and re-pinned. Another two hours wasted, to say nothing of the struggle to do the operation on a reeling deck.

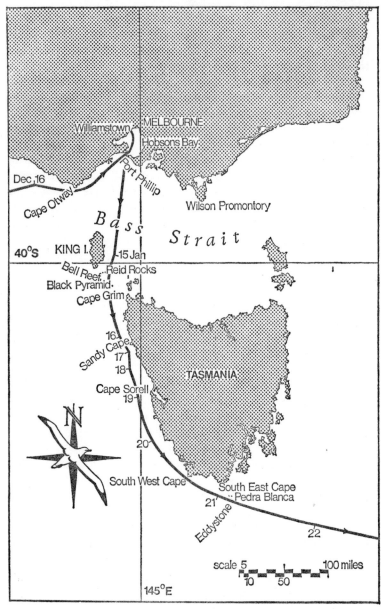

3 Approaches to Melbourne and Tasmania

I made contact with Esperance Bay radio on the 1st December and it cheered me up immensely. We were in the middle of a flat calm that lasted three days. Not a breath of air, not a ripple on the water, but there was lovely warm sun. The deck was a strange colour – it was dry.

Then a breeze got up – but it was south-east. I headed south, close-hauled, but a beat to windward was just what I didn't want. I hoped by getting south the wind would go to west. I spoke to Esperance Bay radio again on December 3rd and got a message from home.

Instead of going west the wind backed to due east and increased to force 5 or 6 with the sea also short and sharp. I was frightened to push her too hard into the sea, close-hauled, as it is hard on the gear at any time, and with my suspect backstays it would have been foolish. My policy, now, was to arrive, and with my mast standing.

I could hear Esperance Bay calling me but my radio couldn't reach them. I ran the engine to charge the batteries, but to no avail. I thought it was the transmitter at fault, but I found out later that it was my batteries which were down. A fuse had gone and although the dynamo was charging it wasn't reaching the batteries.

Those easterlies went on for a week, ending up with a gale from the north-east. We were down to latitude 41°S. and longitude 130°E. The wind then went to north, which at last gave us a good run in the right direction. Then it fell calm until the wind came in from the south with a sudden squall and blew force 8, but we were able to lay our course of ENE. The weather was very changeable and the wind varied from south to south-west and north-west until December 15th, when it blew a fierce gale from the north-west with big seas breaking over us. We were heading for the channel north of King Island (which lies midway in the Bass Strait between Tasmania and Cape Otway, Australia). The Admiralty Pilot warns mariners to beware of strong currents setting on to King Island, a notorious place for wrecks. I had not been able to get a sun sight for a couple of days and I wasn't quite sure of my position.

I was lying a-hull about 120 miles west-north-west of King

Island when a coaster circled me and waved. I was relieved as I knew I would be reported. Since losing contact with Esperance Bay I had been out of touch and I thought the folk ashore might be getting worried. The weather eased and I set spitfire staysail, trysail and storm jib, just as it got dusk. Soon after that I sighted a flashing light to the north-west and identified it as Cape Nelson. I was a little farther north than I had thought, but that was all to the good.

The wind fell very light the next day as I made my way along the coast towards Cape Otway, under working jib and mizzen staysail. An aircraft circled – taking pictures, I presumed. I was nearly becalmed, with sails flapping, but just made steerage way. I came up to Cape Otway at dusk and hauled round on course for Port Phillip Heads. I stowed the mizzen staysail and set the mainsail and crept along through the night, past Split Point lighthouse, and came up to 'The Heads'. Several boats were out to greet me and one – a television launch – came alongside for pictures.

I was warned of the dangerous currents and eddies of the narrow channel through the Heads – but I passed through safely, feeling the swirl as I did so. I didn't know it, at the time, but one of the watchers on the point as I sailed in was Mr Holt, the Prime Minister, who was so tragically drowned a few minutes later.

As I sailed on up through the West Channel of Hobson's Bay more and more craft appeared to escort me and a police launch came out to see fair play. The customs launch came alongside and the officers came on board to clear me. Then to my delight my son, Michael, and his wife, Judy, came alongside in a smart yacht of the Royal Yacht Club of Victoria. They came on board with a bottle of champagne and we drank to the successful completion of the first half of the voyage. Graeme West, the Vice-Commodore of the R.Y.C.V., then arrived on board to pilot me to a berth in their Marina. He took the helm, with the engine running, while I stowed the sails with the help of my son. A great crowd were there to greet me. I was amazed at the interest shown by everyone.

On stepping ashore on the jetty I was greeted by a representative of the Governor-General of Australia and also the

Governor of Victoria, Sir Rohan Delacombe, who read messages of congratulations and greetings. I was also greeted by the Mayor of Williamstown, the port at which I tied up. We then adjourned to the clubhouse for a drink and greetings by the Commodore, John Hayward, and officials before attending a press conference in an adjoining room. I was overwhelmed with the welcome I got, and I was tired but happy by the time I arrived at my son's home in Williamstown for a hot meal, a hot bath and bed.

I O *Melbourne*

F ROM the start everyone in Williamstown was determined
to make my stay as pleasant as possible. The Royal
Yacht Club of Victoria, of which the Duke of Edinburgh
is Admiral in Chief, made me an honorary member and Charles
James, the secretary, was always on hand to help.

One of the first offers of assistance came from Mr Chisholm
Cutts, managing director of Inglis Smith and Co., of Melbourne, who offered to renew any rigging that needed it. This
was a great help and Eric Masey, their sales manager, and Bill
Romas, a rigger, came down and took stock. The two lower
backstays were the first things which needed renewing and I
asked them to add two extra lower shrouds leading to the
backstays fittings on the mast. This involved making up two
new stainless steel plates top and bottom to take the extra
stay. A new bobstay chain was fitted. It was Christmas, which
in Australia is holiday time when firms shut down completely for two or three weeks, but Bill Romas gave up part of
his holiday to work on the rigging. Another firm, Balm Paints
Ltd, came and offered to repaint the yacht completely and
anti-foul her. This I gladly accepted.

Many telegrams were received from friends and relations in
England as well as many in Australia, including the Governor-
General, the Governor of Victoria, and the Lord Mayor of
Melbourne. Then came one from Her Majesty the Queen and
Prince Philip. I shall always treasure that one in particular. I
was inundated with invitations to attend functions and dinners.
My son, Mike, and Judy, his charming wife, acted as my

public relations officers and I don't know what I would have done without them. They kept my engagements book and looked after me all the time. I made friends with Chris and Nigel, my two grandsons, whom I had not met before. On Christmas Day we had a lovely family party at home, starting with early morning service at the parish church.

It was arranged with the Yacht Club that they would slip the yacht on Wednesday, December 27th. This we did without fuss or bother. The bottom was remarkably clean with a sprinkling of large 'goose-necked' barnacles under her stern.

Mike and I spent a pleasant day as guests of the Port Phillip Pilots' Association on board their pilot cutter *Wyuna*. They keep station outside the Heads to transfer Port Phillip pilots to ingoing and from outgoing ships; Captain Alec McAdie was the pilot in charge. Then came New Year's Eve and we attended a party given by Captain and Mrs McAdie in conjunction with Mr and Mrs Swinburne. This was a great night, complete with Scottish pipers at midnight.

In the meantime work was progressing on the yacht. Rod Shepperd, a Balm Paint representative, had undertaken the job of cleaning off the topside and repainting. He was helped voluntarily by Barry Higgins and Alf Watson. Many hours of hard work they put in, and I am grateful to them. Bill Romas came along and set the rigging up, with Eric Masey. The sails were taken off and sent to the sailmaker, who suggested attaching the slides to the mainsail by small stainless steel shackles, to get over this trouble of the seizing chafing through. I agreed and this he did. They lasted all right on the homeward trip.

One night a telephone call came through from England. It was the Lord Mayor of Portsmouth, Councillor D. Connors, to give me the news that it had been unanimously decided to give me the Freedom of the City of Portsmouth. This was wonderful news indeed. This is the greatest honour a city can bestow on one and I was grateful and honoured, but it made me feel very humble really.

Queenscliffe is a small town and fishing port just inside the Heads which claims that the first landings were made there before the settlers got to Williamstown and Melbourne. It was the first harbour I passed after entering the 'Heads'. The

Mayor, Councillor St John, invited me down there to a civic reception. He greeted me on the Town Hall steps and made a speech of welcome to which I replied. After lunch at the Queenscliffe Hotel we visited Lonsdale lighthouse which is on the point at the entrance to the Heads. From that vantage point I could see the dangerous swirls and eddies round the outlying rocks in the entrance and it made me shudder to think of what it would be like trying to pass through in heavy weather.

We were shown round the Army Staff College there. A delightful end to the day was a visit to the Swan Bay Boat Club, whose members, mostly youngsters, in dinghies, staged a sail past in my honour. On the way back we called at the Royal Geelong Yacht Club and I was presented with a burgee as a memento.

A pleasant morning was spent in the market place, where Mr Frank Nurse, secretary of the Retail Fruit Traders Association, introduced me to members. They later presented me with all my fresh fruit and vegetables for the return trip.

A memorable occasion was to be received by the Lord Mayor of Melbourne, Councillor Talbot, at a civic reception at the Guildhall, where I met many local personalities.

This was the pattern of my stay in Australia. There just were not enough hours in the day to fit everything in, and much as I would have liked to, I couldn't afford the time to stay too long. I had to fix a date for leaving and arranged for Sunday January 14th. I spent as much time as I could doing odd jobs on the yacht. Club members volunteered to help. The 'Tiny Tim' charging engine had seized up. I got that out and a club member took it away and got it going. The quadrant on the servo box of the self-steering was 'working' and another club member took it away and bolted stainless steel brackets to hold it firm; he also made up some more swivel pins of stainless steel and fitted them, and gave me a box full of spares. Graeme West obtained two new servo-blades as spares, and changed the main engine oil for me. Albert Klestadt was always on hand with offers of assistance and helped in various ways enormously, with his two sons Geoff and Jonathan. Albert had made a remarkable voyage in an open boat during the war when escaping from the Philippine Islands on the approach of the Japanese. With only a school atlas as a chart he sailed, with a native crew, across open sea, to Northern Australia. He described his experiences in his book *The Sea Was Kind*, a copy of which he presented to me.

We were entertained at the Lido Night Club by Eric and Janet Smith of H. J. Heinz and Co., who kindly offered to stock me up with tinned foods. Other gifts of stores I received included: specially selected eggs from Victorian Egg Board; tinned milk and cream, Tongola Milk Products; petrol, oil, and paraffin, Ampol Petroleum; tinned chicken, etc., Shippams; Plymouth Gin, Seager Evans; fresh fruit and vegetables, Retail Fruit Traders Association; specially baked bread, Bates Bakery, Williamstown. Many individuals also came forward with gifts of foodstuffs and drink. To all these kind people I wish once again to express my grateful thanks.

Mike and Judy stowed the stores in the various lockers and listed them in my stores book.

On the Friday before I left I was entertained at a cocktail party at the R.Y.C.V. Here the mayor of Williamstown presented me with a wonderful bronze plaque, specially cast, commemorating my voyage to Australia.

On Saturday we were ready for launching. *Lively Lady* had been anti-fouled by a special preparation made up solely for the purpose, which had been mixed at Sydney, at the firm's works, by their chemist Peter Wright. It proved its worth.

Just before we launched I was honoured by a visit from Sir Rohan Delacombe, Governor of Victoria. He came on board and was very interested in everything – especially the self-steering gear.

Sunday morning came and I woke with a heavy heart at the thought of leaving my son and his wife, Judy, who had looked after me so well, and my two grandsons. Not only them, but a host of others who had endeared themselves to me by many kind acts, brushing my words of thanks aside with such remarks as 'Only too pleased to help'. I found the same hospitable welcome amongst all Australians wherever I went. I was impressed by their regard for England and their love for our Queen. Her picture hung in prominent places from the

R—G

Guildhall to the local hostelry, and their loyalty cannot be questioned.

We attended early morning service at the parish church, Holy Trinity, and I was moved when the priest said a prayer for those about to embark on long journeys. I could hardly speak when he shook hands with me afterwards and wished me God-speed.

Down at the Yacht Club, Graeme had moved *Lively Lady* round to the outer jetty and she was straining at her mooring ropes as though anxious to be off. A special delivery telegram arrived for me from the Governor-General of Australia wishing me well from all Australians. Then I was called to the telephone and it was a call from Dorothy.

A large crowd had gathered to see me off, and a tug lay nearby with the local television cameras installed. Tony Charlton, the commentator, and I had become friends since meeting on my arrival. The Mayor of Williamstown and officers of the R.Y.C.V. were there, and as we cast off a great cheer came from the crowd and ships blew their sirens. Graeme West piloted us into the channel, then left us. Mike and Judy were accompanying me as far as Queenscliffe, just inside the Heads. A host of small craft were escorting us, including some from the Royal Melbourne Yacht Squadron, who had made me an honorary member.

It was a fresh southerly wind which headed us so we were forced to tack down Hobson's Bay. It brought short seas and made it wet, cold, and uncomfortable. One by one the small craft dropped away, until only the television tug remained. Then she gave a toot and turned away and left. We continued, tacking down the narrow West Channel of the maze of sandbanks at the southern end of Hobson's Bay. Time was pressing if I was to clear the Heads before the tide turned at dusk, so I put the engine on. It was late afternoon when we were off Queenscliffe and a launch appeared with the Mayor and several councillors on board to wave me off. Mike and Judy had planned to board her but conditions were too rough for her to come alongside, and so they transferred to a small power boat. I said goodbye to them with a lump in my throat. I had not met Judy until I arrived in Australia and we had come to love each other dearly. They had done everything they could to make my stay a happy one and all I could say was:

'Thank you, it has been one of the happiest months of my life.'

They turned away for Queenscliffe and I was alone, heading for the narrow channel and through the Heads. A large white-hulled passenger ship came down, outward bound, and gave me three loud blasts on her siren. Her pilot must have been a sailing man as he came right round my stern to avoid passing to windward of me and taking my wind.

I passed out through the Heads, at dusk, with the Lonsdale lighthouse flashing out her warning light against those treacherous rocks. My thoughts flashed back to Captain Alec McAdie's last words to me as we shook hands on my departure. 'You will come back, won't you? Promise?'

I I *Bluff for Repairs*

For maps see pages 64, 76 and 93.

O UTSIDE the Heads the seas were bigger, of course, as we headed south-west against a stiff, southerly wind. The motion was severe, and *Lively Lady* was throwing water and spray all over as we bashed into the seas. I kept watch as there were lights of several steamers about. Last-minute gifts and gear not properly stowed were flung across the cabin. I could have done without this severe thresh to windward at this stage of the voyage. The pilot cutter was on station, transferring a pilot to a large tanker; then she steamed over to me and hailed 'Good luck, *Lively Lady*'. This was a heart-warming gesture.

At midnight I called up Lonsdale lighthouse, and spoke to the keeper. Since my visit there we had become firm friends. He, having sailed twice round Cape Horn in square riggers, gave me the course they had steered and their daily runs. This was interesting to me, as I was able to compare relative conditions on my way round.

My first morning at sea found me still fighting to windward. It had been a rough night, with heavy water on deck. The light genoa, which had been returned from the sailmaker, had been left bagged up on deck, and was missing, washed overboard. I was cross with myself for not lashing it down; but it was too late for regrets now. It was gone.

By mid-morning the sun was out, the sea and wind down.

The wind hauled round more to the east, so I decided to shoot through the passage between King Island and Tasmania, and altered course more to the south. Soon after that, I sighted

King Island to the west. That evening I spoke again to Lonsdale lighthouse, and also to Melbourne radio.

At dusk, Black Pyramid Rock was abeam. This is a big, black unlit rock sticking up in the channel off Tasmania, and I was glad I passed it before dark. The wind was easterly and light, and I made slow progress through the night. By midday on Tuesday, January 16th, the wind had backed from east right round to south-west, keeping me busy with sail changes all the time. It was very light and we only just made steerage way. The wind went to south, and it was tacking and tacking again off Sandy Point, Tasmania.

The smart fishing ketch *Julie Burgess*, came alongside and threw me over a sack containing some crayfish. He asked to be remembered to Alan Villiers.

That evening when I spoke to Lonsdale lighthouse and Melbourne radio, they passed a message from the Royal Australian Naval Sailing Association, Sydney, wishing me luck. I needed it, for that night a strong gale blew up from the south. I stowed the working jib and lay under reefed mainsail, heading west-south-west offshore. I was dead tired, and turned in for a couple of hours. Running down the coast is worrying enough without a gale from dead ahead.

In the morning I set the spitfire staysail and went about on the other tack and closed the land again. I was south of Sandy Cape on the west coast of Tasmania, so I had gained a few miles. By evening it was easier and I took advantage of the conditions to have a good sleep. I hadn't had much since leaving, and I was tired and stiff with the violent motion. The morning of January 18th was sunny and calm. After the gale against us, now it was flat calm, with what wisp of air there was coming from the south.

It was tack and tack to gain a few miles south. The wind went right round the compass twice, and I was kept busy changing sail to try to make the most of it and coax the yacht along. I lowered the mainsail and set the mizzen staysail. I lowered the working jib and set the big genoa. Then the wind went to south and I lowered the mizzen staysail and set the mainsail, lowered the big genoa and set the working jib. So it went on. Whichever tack I went on was the wrong one. By the fifth day

I was becalmed off Cape Sorrell, Tasmania, having covered only 330 miles in five days. We ghosted along, getting south.

I got the anchor stowed below deck, and re-stowed some of the stores to lighten the yacht forward. I turned out some old, rusty tins and dumped them overboard. By the 21st – seven days out – I passed Port Davey Light, and with a south-westerly breeze headed south-east past South West Cape, then South East Cape across the south of Tasmania. I sailed past Pedra Blanca Rock, a favourite breeding ground for gannets; then the Eddystone Rock, which stands up stark and straight like a lighthouse. These rocks lie some fourteen miles off the south Tasmanian coast and, with no light to mark them, it would be certain destruction to hit them on a dark night. I wondered how many have done so in the past.

I called Hobart radio, but it was impossible to contact them. The fishing boat *James Lee* called me up and relayed my position. They said they often had difficulty in contacting Hobart.

Later that night I made contact with Hobart myself, and passed a message to the Governor-General, thanking him for his telegram and, through him, all Australians for their hospitality. The wind went to west, very light. I lowered the working jib for boomed out genoa and lowered the mainsail and set mizzen staysail. The wind went to north-west, and I changed the genoa to the starboard side. The wind increased and with the mizzen staysail the yacht was yawing about unbalanced, so I lowered it. We were getting along well, at six knots, on course to clear southern New Zealand. I set the genoa staysail on the little boom, port side. Then the wind went to south-west, strong, and I had to lower big genoa on starboard side. I hoisted mainsail, stowed genoa staysail and set working jib on boom port side. All this during a few hours, and people say, 'What do you do with yourself?' I counted up that made sixteen sail changes that day.

The night of January 22nd/23rd was calm, and I had a good sleep, which I needed. I awoke, to find *Lively Lady* heading south, not that we had done many miles. The wind came in from the north-east and we were soon close hauled. It gradually increased and by afternoon it was blowing a gale.

I shortened sail and by midnight I wondered if I ought to lower the reefed mainsail; but we were bashing on in the right direction and at dawn it eased. I set the working jib and took the reefs out of the mainsail. The wind fell very light and thick fog came down, with visibility no more than a hundred yards. I heard a plane overhead, and he circled, but I couldn't see him. I wondered if he had picked me up on his radar; but he soon faded away.

By midnight of January 24th/25th the wind was freshening from the north. We were getting along well, but it increased to force 7 and I judged it time to reduce sail again. It was pitch dark, heavy rain and brilliant lightning. I went forward and let go the jib halyard. It went with a rush and nearly took me with it. The whole lot – jib and the twin forestays – all came down and fell over the side, into the sea.

The stainless steel fitting at the masthead had parted. The top of the mast was bending back dangerously and I let go the main halyard and clawed the mainsail down and stowed it. I then fished the jib out of the sea, together with the twin forestays, and there, on the end, was the broken tang of the masthead fitting. I unshackled the jib halyard and fastened it down to the bowsprit and winched it up as forward support for the mast. I then retired below to a hot drink, and turned in for a couple of hours, to consider what to do. I was shattered and stunned at this piece of bad luck. Who could have foreseen this?

At dawn I surveyed the damage and cleared up on deck. I took the jib off the forestay, and unshackled the two forestays, stowing them away. I thought I could shackle the jib on the one halyard that was shackled down to act as a forestay and hoist it with the other halyard. It was raining, with thick fog, while I did this; but soon after it fell a flat calm. I was feeling very depressed and miserable, and to crown it all, I had a touch of lumbago. The calm lasted all the next night, and I turned in for a rest.

At dawn on January 26th my back was better; a light breeze came from the north-west and I hoisted the working jib and mizzen staysail. There was a big swell, causing us to roll heavily. The wind went to south-west and I went over to the other tack.

I then ran the charging engine, just to dry it out, and cleaned the plugs of the main engine. I had decided to make for Bluff in southern New Zealand. Hobart was a little nearer, but Bluff was down wind and on my way.

That night I called up Arawua radio, the station for southern New Zealand, and was surprised to get them straight away. I had a message from the *Sunday Mirror*, asking for an estimated time of arrival at the Horn. What a joke, I thought, in my present condition. The wind fell away and all next day we rolled round with sails flapping, but making a few miles in the right direction.

On the night of January 27th/28th I spoke to Arawua radio again, and Melbourne radio came in at over eight hundred miles. I was pleased to hear them, and asked to be remembered to all my friends in Melbourne. I asked Arawua radio to contact the harbourmaster at Bluff, to ask if he could give me a berth alongside, where repairs could be made to the masthead; or alongside a big ship, as I anticipated then that small-power tools, such as an electric drill, would have to be used at the masthead. I also asked that a piece of stainless steel be acquired, ready to strap over the top, to take the forestay fitting. I did not then know that my friend, Sir David Mackworth, was arranging to fly out with a complete new fitting.

I passed a lengthy message to the Press, and ITN in London, but had to break off in the middle to lower the mizzen staysail. The wind had increased and the yacht was off course. This mizzen staysail certainly wants watching, and is not a sail to leave up while turning in for a good sleep. So I left it down, had a couple of hours' sleep, then reset it, with the wind light northerly. By noon the wind had increased to force 5 and I lowered the mizzen staysail and set the trysail. It was cold, with misty rain, and as the wind increased so the sea got up with dollops coming aboard. The jib halyard, to which the jib was shackled, parted at the rope tail which was round the drum of the winch. I lowered the jib and set the spitfire staysail as the wind increased to gale, and suddenly went right round to south.

It was a wild, rough night. I made the remaining jib halyard fast to the bowsprit, and ran all night under staysail and trysail.

I contacted Arawua radio again and received messages from home and also from the harbourmaster of Bluff, advising me not to attempt to enter Bluff harbour on my own, but to wait at the entrance for him to pilot me in. The weather eased next day, and I changed the spitfire for the genoa staysail; but by noon on January 30th it increased again to gale, and I changed back to spitfire. It got really bad and, in the end, I was forced to lower everything and lay a-hull.

The wind shrieked and tremendous waves rolled up, some hitting us hard and coming over into the cockpit. It was rugged. The gale was from the north-west, driving us south-east. I was heading for the narrow passage between South Island of New Zealand and Stewart Island. This island is unlit and, with no sun sights for a couple of days, I was on dead reckoning for my position, and could be a few miles out. Care was necessary. The gale backed to south-west and blew with renewed fury. It lasted 24 hours, blowing up to force 10. We were knocked about very severely, being rolled well down to leeward, and big seas hitting hard and breaking over us.

By noon on the 31st it eased enough for me to set the spitfire staysail. I plotted my assumed position which put me just south of Puysegur Point lighthouse, on the south-western corner of South Island. I had run some 1300 miles from Melbourne in seventeen days. A slow passage. I still had nearly a hundred miles to go before reaching Bluff Harbour. I set a course to leave Centre Island to port. This is a small island off the rocky coast of the mainland of South Island and between that and Stewart Island.

It was a murky, overcast day, with poor visibility in the mist; but just before dusk I sighted an island standing up out of the water, black and forbidding. It could only be Solander Island; a big, rocky island dead in the middle of the channel between South Island and Stewart Island. It is unlit, so I was glad I had sighted it before dark. It was just where I wanted it, on the starboard bow, and I was on course for Centre Island. Just before dawn I picked up Centre Island light, just off the mainland and on the port bow. The wind eased and went due east – right on our nose.

I set the genoa staysail in place of the spitfire and also set the trysail. I was afraid to set the mainsail with no proper forestay. It was all tacking in the narrow channel. I radioed the Bluff harbourmaster that I reckoned on being down to the harbour entrance by about 5.30 P.M. However, the wind went round to south-west and freshened. I lowered the trysail. I prepared for entering harbour. I got the anchor up from below deck and shackled on to the chain. I got fenders and mooring ropes ready.

By this time the wind was strong and we were making a fast passage down to Dog Island lighthouse at the harbour entrance. So fast indeed that I had to lower sail and wait for the pilot boat to arrive. The wind increased to gale force and the sea became a mass of white foam, as the pilot boat appeared and piloted me into the narrow channels of the harbour entrance. A fierce tide was running, and I could see why the harbourmaster had advised me to wait for his assistance to enter.

I was rather pleased with myself over my navigation. It had been three days since I had had a sun sight to fix my position. I had had a severe gale during which the drift must have been considerable; yet I picked up my landmarks, in a narrow, rock-strewn channel, just where I expected or hoped to find them. I had been amazingly lucky.

As we cruised up the harbour and along the headland, I could see rows of cars parked along the coast road. I wondered if it was a half-holiday or some special occasion, but as I passed them they all set off, heading back towards the harbour, and I began to realize they were there to watch me come in. As I got farther up I passed ships which started sounding their sirens and, in particular, a dredger which made more noise than any. I was shown to a landing stage where willing hands took my mooring ropes. A crowd was gathered to welcome me, headed by the Mayor, Councillor Johns. The Press, radio and television cameras were there, and it was a right royal welcome. It was then that I learned that Sir David Mackworth was on his way by air with a new masthead fitting. I couldn't believe it at first, but I was assured that it was true.

The hospitality of the people of Bluff was soon apparent.

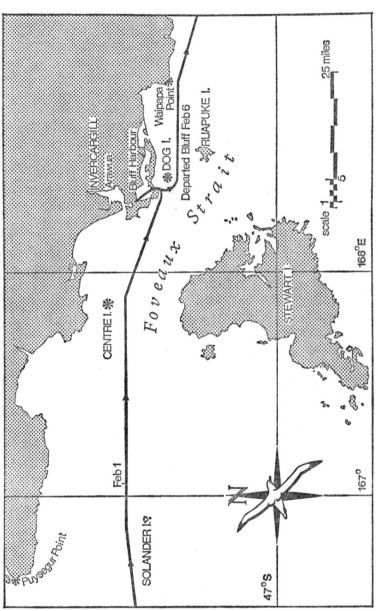

4 Approaches to Bluff

A gift of fresh fruit was made to me by the Fruit Federation. The Mayor then informed me that accommodation had been reserved for me at the Club Hotel, a leading hotel in the town. They assured me that the yacht would be watched all night, and would be quite safe. I gathered my bag together and the deputy Mayor, Councillor F. Dawson, drove me up to the hotel, where I had a hot meal, a bath, and to bed.

The next morning I was driven down to the harbour. *Lively Lady* had been moved round to a dock basin. It had blown hard during the night and her rubbing strake had been damaged. This was put right very efficiently, however, and cannot be faulted. I was introduced to the managing director of the Bluff Engineering Company, Mr Lewis, who very kindly offered to undertake the repairs of the yacht at no charge to myself. This was a wonderful gesture of goodwill, and I was deeply grateful. David was due at about noon, and we put off any decisions until he arrived with the new fitting. In the meantime I was cleared by the Customs Officer, and introduced to the manager of New Zealand stevedores. He kindly offered to replace my broken halyard, and to do any other rigging work required.

David arrived and we got down to working out the details of the repairs. I hoped it could be done without taking out the mast. Mr Lewis put two good lads on the job: Jack Crooks, who had built his own yacht, and Peter Eastlake. Jack was hoisted to the masthead, and said he could do it without taking the mast out. This meant taking the topmast shrouds off and the topmast backstay, to remove the old masthead fitting; the mast being supported by the lower forestays, lower shrouds, and lower backstays. The new masthead fitting was then fitted and bolted through and the topmast shrouds and backstay connected and set up. I was thankful when it was done. Work proceeded well. In the meantime Shell Oil came and offered to top me up with petrol and paraffin.

One evening, after dinner at the hotel, I had a visitor in the shape of Mr Stephenson, the manager of the Invercargill branch of the Bank of New South Wales. He had come to offer me financial assistance, should I require it. I was able to tell him that the hospitality I had received was such that I did not need any. It was an extremely kind and generous gesture. Before I left he presented me with a beautiful book on New Zealand, which I shall always treasure.

During my voyage round the world I was greatly impressed by the friendliness and help I received from the officers manning the shore radio stations. They were always so courteous, patient, and helpful; but none more so than the officers of Arawua radio of New Zealand. Three of these lads came to see me when I arrived, and were interested to see my radio installation. One evening the Mayor drove us up to the radio station, where we were welcomed and shown round by the principal radio operator. He had taken the trouble to type out full instructions on contacting them from extreme limits. They couldn't have been more friendly or helpful and this warmth came through to me over the air. I would like to say 'Thank you' once again to these lads.

We were constantly being invited out and spent most evenings in this pleasant way. One couldn't imagine meeting more hospitable or kind people. At last on Monday evening we were finished – four days after getting in. The masthead fitting was in place, the stays set up, new jib halyards and a new main halyard, water tanks emptied and refilled and fuel tanks topped up.

On Tuesday morning I was ready. David was to return to England, and had to get to Invercargill to catch his plane. On the dockside I received many small gifts of foodstuffs and drink from individuals too numerous to mention, but ranging from a jar of home-made blackcurrant jam to home-made biscuits, whisky, and a case of canned beer from the workers of Deep Cove.

I cast off from the dockside and moved round to the landing stage, where the Mayor was waiting to give me an official send-off. A big crowd was in attendance as well, including children from the local school, who lined up. Two of the seniors, a boy and a girl, read an address to me and presented me with their emblems. It was all very intimate and friendly and after the Mayor had said his piece and had shaken hands with me I was too moved to reply. The Mayor and I had become firm friends, as well as many others present. I cast off and followed Captain Bird, the harbourmaster, out towards the channel to the open sea. There was a great cheer, car horns sounded, as well as ships' sirens. Fireworks and rockets were sent up, and two big Blue Star ships, which were in the harbour, dipped their ensigns to me. Captain Bird escorted me out as far as Dog Island lighthouse, the outermost point of the harbour. I set the working jib and, with a farewell wave, I was on my way. The pilot launch turned back, and my last sight of her was of Mr Lewis, Jack and Peter who had accompanied Captain Bird, waving both arms in farewell.

I was alone again, but greatly moved by the warmth and kindness of these wonderful people.

1 2 *Roaring Forties*
For maps see pages 64 and 103

IT was blowing force 6 from the west as I headed east on February 6th after waving goodbye to the pilot launch. After five days in harbour I was headed out through a narrow channel, the Foveaux Strait, only five or six miles wide, with jagged outlying rocks on either side. Mist blotted out the coast on either side. By the late afternoon it was blowing up and I changed the working jib for the spitfire staysail and ran on at about five knots. At dusk Waipapa Point light on the eastern corner of South Island, was bearing north, and I was clear of that channel and heading into the open sea, 5,000 miles of it, to Cape Horn.

It blew up to force 8 during the night, which was cold and rough. It is certainly a windy spot down there. It blew up as I arrived, and as I left, as well as most of the time I was there. The Foveaux Strait, which I had just passed through, is not a nice area to be in in rough weather and the Admiralty charts warn against strong currents and unpredictable magnetic compass variations; which means one cannot always rely on the compass bearing. I was glad, therefore, to be in the open sea again.

The next morning was clear, with the wind west-north-west and lighter, and I set the big genoa on the boom and the mizzen staysail, but by the afternoon it veered to north and then north-east. I lowered the mizzen and set the mainsail, and lowered the big genoa and set the working jib. We were close hauled and soon after that I had to reef the mainsail.

That night I spoke to my friends at Arawua radio and passed

on messages of thanks to the Mayor and through him to all my friends in Bluff. The next day we were almost becalmed. I stowed away last-minute gifts and generally tidied up down below; the mooring ropes were coiled up and stowed below, as were the fenders. I unshackled the anchor and put that below deck. Then the wind came in from the east and by midnight was blowing up to gale force and I had to shorten sail. The sea was short and sharp and we could make no headway against it, without laying her off the wind and heading almost south. I noticed the wind vane was behaving in an erratic manner and found the bolts, going through the stainless steel vaneshaft it is mounted on, had sheered. I repaired this with new bolts. I foolishly tried to do it without putting oilskins on and got wet. My own fault.

It was rough, heading to windward with *Lively Lady* burying her lee rail and throwing spray all over. The glass continued to drop. Suddenly the wind fell to calm and left the sails slatting about. Just as suddenly a fierce squall descended on us from the west and knocked us almost flat. I had to scramble forward and lower everything. What an awful night that was; the wind screaming and great waves hitting us, laying us well down, with white foam across the deck. Down below it was difficult to move about, such was the violent motion. Anything not properly stowed was thrown across the cabin. I made things as secure as possible and did the only thing – turned in.

The morning of February 10th dawned clear but the wind was still shrieking. I tried to get moving under spitfire staysail, but the seas caught us under the stern and swung us right round and she would not come back. At last, by mid-morning, it eased enough to get along under spitfire staysail, but she rolled horribly, scooping up water over the side decks.

I had a late breakfast and lunch together that day: porridge, with brown sugar, raisins, and milk, followed by fried potatoes, tomatoes, egg, and sausages, and a fresh apple, washed down by a cup of coffee.

The previous night the radio telephone had been very weak. I tested the batteries but found them well up. On looking round the connection I found the lead-in from the aerial broken. I repaired this so I hoped it was now all right. I set the working jib on the boom, but as the wind went to north-west, I had to stow the boom and raise the reefed mainsail. That night I got Arawua radio successfully and received messages from Dorothy and David.

We were now down to 48° 10′ S. and 178° 40′ E. It was cold, with a grey boisterous sea. We were getting near the northern limit for icebergs. The sea is cruel down here; it comes at you in a wicked way and is out to get you. Even in moderate conditions it has the habit of bringing along that awkward cross sea, which throws us about and slops on board and into the cockpit, making it impossible to go outside at all without oilskins on. Everything is cold and damp below. However, I managed to keep myself dry underneath my oilskins, and so to keep warm. I kept my bunk dry also, so was always able to pull dry blankets over me.

On Sunday, February 11th, I crossed the dateline by the chart, so I had two Sundays, the 11th and the next day. I mention 'by the chart', as the International Zone time is actually a few degrees East. However, to save complications, I altered my time as I crossed longitude 180°. I should explain that the international dateline is in principle longitude 180 degrees east or 180 degrees west of Greenwich, which are the same thing and on the opposite side of the world to Greenwich. Actually it is modified a bit ashore as the longitude line passes through various groups of Pacific islands, and it would be very awkward if it was Tuesday on one island and Wednesday on another island a few miles away.

When travelling eastward from Greenwich, an hour must be added to the time for every 15 degrees of longitude, so that the sun continues to be at his highest around midday by the clock; this goes on until 12 hours has been added when approaching the dateline. Similarly when travelling west one hour is deducted from Greenwich time for each 15 degrees of longitude, until 12 hours has been lost on coming up to the dateline. I was travelling east so I apparently gained a whole day on the dateline to get two Sundays, but actually this was cancelled out, because I was losing it in 24 instalments of an hour each all round the world.

R—H

The 12th was a muddling day; in the early hours of the morning we were getting along at about six knots, under reefed main, spitfire and storm jib when the northerly wind increased and I dressed in oilskins ready to shorten sail. It was rough going, *Lively Lady* burying her lee rail as a beam sea caught her, but we were heading in the right direction. Suddenly the wind dropped right off and went round to south-east. I changed the storm jib for the working jib and the spitfire for the genoa staysail. I noticed a seam undone in the leach of the mainsail so I lowered it and sewed it up. On trying to set it again, I found the halyard had got round the foreside of the mast, at the top, and wound round the navigation lamp. Try as I would, I could not free it. In the end, I took it off the sail, shackled a thick rope on to it and hoisted that. It freed itself and I was able to hoist the mainsail again. The dodgers were torn away at the eyelit holes and I threaded some nylon cord through and seized them up again.

We were now close hauled again with a south-easter. We made good progress and noon-to-noon run was 148 miles. It was rough going, though, bashing to windward, hitting the seas hard and throwing water everywhere. Grey low clouds swept by, with mist and heavy rain. One needed three arms to hold on with below. I watched from the hatchway, oilskins on, ready to shorten sail. I was quite out of breath with the tenseness of the situation. In the end I had to lower the genoa staysail, change the working jib for the storm jib and reef the mainsail, as the wind increased to force 8 in gusts. We were still doing six knots. That night I spoke to Arawua radio at 850 miles. I sent a birthday message to my wife for February 15th. During the night the wind dropped away to nothing and the sails were slamming. I had to dress up in oilskins to go outside to harden in the sheets in heavy rain. By dawn on February 14th it was blowing hard again from the south-south-east, and as it went to south it increased to force 9. I lowered the mainsail and storm jib and set the spitfire staysail, with solid water coming over me, into my eyes and nose. As the afternoon wore on it increased to storm force 12. I lowered the spitfire staysail and set a baby storm jib. This sail just prevented the pounding under the counter. The wind shrieked and howled, and the

hail and rain tore at my face and blinded me. I wrote in my
log:

> This must be a Southerly Buster, never have I had such a fierce
> wind, its fury shakes the whole ship. Big seas constantly sweep the
> deck and cockpit. How the wind vane stands up I don't know. All
> I can do is to watch, oilskins on, in apprehension.

I felt weak and dizzy with the motion, and made my usual hot
drink to pull myself together.

At dawn on February 15th the wind eased. After a cup of
tea I looked round on deck; the starboard dodger was badly
torn and the iron stanchion which carries the guard rail was
bent inboard. It must have received a heavy blow to do that to
it. The wind went to west for a few hours, enabling me to run
with boomed-out genoa one side and working jib the other.
But the next day back it came to east, with us close hauled under
all working sail, ending up with a strong gale on the night of
the 17th. Never have I seen the glass go up and down so quickly
– particularly down. The seas were terrible and we were laid
down so far that a case of lemons in the fore-peak was turned
right over.

The night of the 18th was one I shall never forget, as it
nearly put paid to me and the trip. I had started the little Tiny
Tim to top up the batteries. This little engine, being slung in
gimbals, has a flexible exhaust pipe. I disconnect this when not
in use as it always seems to have water in it which gets into the
engine. Apparently I had not connected it up properly because,
on coming down into the cabin, I noticed it smelling of fumes.
I stopped the engine, and it was lucky I did, because the next
thing I knew was struggling back to consciousness, slumped over
the stove. My fingers were seized around the companion steps;
it was dark, and rain and spray were coming down the open
hatch on to my bare head. I was cold, weak, and dizzy and just
struggled to my bunk and lay down, in my oilskins. I had
been 'out' for about three hours, overcome by the fumes from
the engine. I did not feel myself going dizzy or anything like
that – just went out like a light. It was a good job the hatch was
open to clear the air below.

By dawn on February 19th it was blowing a gale and I had

to turn out and lower the mainsail, but I was too weak even to think straight. I had a dose of Andrews Liver Salts and a cup of tea and began to pull myself together, though I did not want anything to eat. The gale lasted all that day and into the next night, before moderating, but it was still easterly and *Lively Lady* took severe punishment driving against it. With the help of a hot lemon and whisky I raised the energy to set the mainsail and we lay close hauled. The wind increased again and the yacht was being so punished that I lowered the mainsail again. Every time the wind tended to go south I hoped to be able to sail more free, but then back it would come to the east again and blow up to a gale, as it did on the 21st. On that morning I wrote in my log:

> Still don't fancy my food and have to force myself to eat. Thank God, I'm feeling a bit better this morning and able to face the deck work in these rigorous conditions. The last couple of days I've been really low and felt weak and ill. Makes one realize the importance of keeping fit.

It blew a strong gale all day, but we pointed north-east, under spitfire staysail. The yacht took some severe knocks, though, laying her well down at times. There were one or two awful lurches when she felt at though she was never coming back. A crash and a plate jumped out of its stowage and broke into pieces.

On February 22nd the sky cleared a little and showed a brighter look about it. The wind was still strong from the east, but at noon the sun broke through and I got a hazy sight, the first for five days. It put us farther north than I had hoped and our position was 43° 40′ S., 158° 0′ W. I had lost contact with Arawua and also Auckland but now made contact with Chatham Islands' radio. I sent my position through to London.

I had had these easterlies for nearly three weeks with constant gales and had been pushed some 300 miles north. The yacht and I had taken severe punishment and I was feeling bruised and sore and depressed at the slow rate of progress. However, on the evening of the 23rd the wind went round to the south and we were getting along well at six knots under all working sail. The wind was gusting force 6 in rain squalls, so I lowered

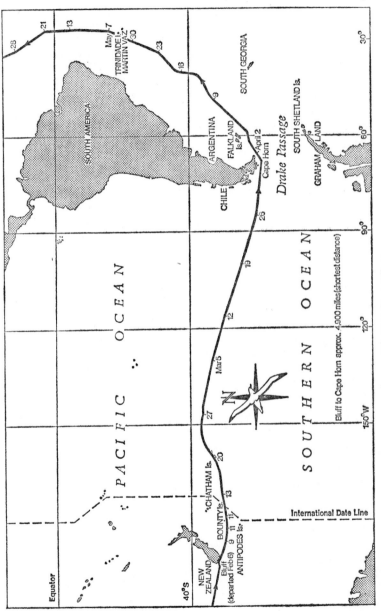

5 Bluff to Cape Horn

the genoa staysail. I heard Chatham Islands calling me on the radio but as I could not reach them I assumed I had lost contact; however they came through again later and picked me up. I believe Arawua radio was listening in as well. The wind continued strong from the south and I sat and watched all night, in oilskins, wondering whether to take the mainsail off, but we were heading in the right direction and it was only in squalls that we were laid well down. We reeled off 70 miles in ten hours. But the sea beat me in the end, as it is determined to do, coming at me on the beam and swinging her round, at times almost out of control. I was forced to lower the mainsail and set the trysail, also the spitfire staysail.

During one rough period, when *Lively Lady* was heading into the seas and taking solid water over the foredeck, I was in the forepeak sorting out some fruit; I could hear water splashing and when I lifted the flap leading to the chain locker I was surprised to see the amount of water that was pouring down through the hawse-pipe. Once or twice lately I had been surprised and a little worried at the extra pumping I had had to do to clear the bilge. From the usual half-dozen or so strokes of the pump, once a day, it had increased to about twenty. Water through the hawse-pipe would account for that. So when the wind eased and the sea went down a bit, I spent some time on the foredeck plugging the hawse-pipe around the anchor chain with rag and tallow. This made all the difference. I also noticed the bobstay swinging about rather, so I hung over the stem and tightened that. The lower backstays that were renewed in Australia had, apparently, stretched a little, so I set them up. There is always work to be done of some sort, maintaining gear on a long voyage such as this.

On the night of the 24th Chatham Islands radio came in on the radio telephone. I made contact and passed my position on to London. This was a good performance on the part of my radio as we were over 1,100 miles from Chatham Islands. It was to be my last contact by radio until Cape Horn.

At Bluff we had replaced the self-steering ropes leading to the tiller with flexible wire with rope tails. We thought these would last better as I had had trouble with broken steering ropes through chafe. However, on the 25th, nearly three weeks out from Bluff, the first steering wire parted, at the bight around the sheave. It was blowing hard at the time and we were getting along well under a working jib with the wind dead aft. I tensioned the tiller with a piece of shock cord and replaced the broken wire with rope. I noticed the other wire was stranded so I replaced that also. It was uncomfortable work, on a heaving counter aft, disconnecting the wire from the quadrant on top of the self-steering gear and replacing it with rope. It was blowing up to gale force and the yacht was yawing about wildly with seas on the quarter. I lowered the working jib and set the spitfire staysail. She now went easier, but just as fast. I tried for a sun sight, but got the sextant wet with spray and had to rinse it in fresh water and dry it.

13 Rounding The Horn

For maps see pages 103 and 152

THE bad weather continued until the night of the 26th/ 27th, when it increased to storm force. I was awakened from a sleep to find her a-back. I rushed on deck to free the wind vane. The wind was screaming and I had to lower the spitfire staysail. I hoisted the baby jib. The seas were tremendous, some breaking right over us. The other dodger was torn away and the heavy iron stanchion that side was bent inboard.

I retired below, and barely had time to put the washboards up when a huge sea came aboard over the port quarter. It slammed straight over the open hatch under the spray hood. A great deluge came below, soaking everything, including the chart table with the open chart on it. I looked out. Everything seemed all right, but on closer examination I found the spray hood had been split open and a Mae West life jacket with a coil of terylene rope had disappeared over the side. The Walker's Log on the quarter had been twisted right round.

I thought the wind had eased a bit, but I have noticed before that after easing, the sea has a last violent fling, just to let us know he is still there.

I mopped up and cleared up below and the next day, when the weather really did moderate, I had a go at repairing the spray hood. I had already changed the baby jib for the storm jib and the trysail. After a good breakfast of porridge, two fried eggs with potatoes and tomatoes, and a hot cup of tea, I felt better able to work. Just before leaving Melbourne a wellwisher had presented me with a roll of 'Sylglass' sticky tape. This is messy and sticky but it will stick to almost anything and is waterproof. It is invaluable for making temporary repairs and I can recommend it as a useful item to carry on board. I stuck a piece of this right along the open seam of the spray hood, holding it together. The spray hood was of a shiny plastic material and I wondered whether it would stick, but it did, and held until I got home.

Day followed day on much the same pattern. A calm followed by a gale followed by a calm and so on in succession. During a calm I would busy myself sorting over the fruit. This was lasting very well and not much had to be thrown away. I made up spare lengths of shock cord to help the self-steering gear hold the tiller. But all the time, in a calm, I would be watching out for the faintest breeze and would trim the sails to coax a little wind into them. It was a blow when one day I found the starboard booming-out pole had broken away from its slide on the mast. It had gone at the welded joint, which seems a weak point on this fitting – a nuisance because it meant I couldn't raise a boomed-out headsail to starboard.

The calms were the most frustrating periods. After a day or two of good runs a calm would set us right back on our daily average. For instance, in the early hours of March 9th it blew a force 9 gale and we got along well under spitfire staysail, but by noon it was a flat calm with the wind vane going round in circles. This lasted 24 hours but by the next night it blew up to force 10 and I was lowering sails in a hailstorm which was lashing at my face. Big beam seas swept the deck. An entry in my log reads: 'Lower working jib in a hail squall, seas bigger and hitting hard. This Southern Ocean seems eternal and rough violent seas and howling winds almost a way of life now.'

I peeled some potatoes and carrots and boiled them for supper with a tin of corned beef, but it was a wicked night, with wind increasing to force 12. One heavy sea came over and I was thrown across the cockpit on to the guard-rail stanchion and received a heavy blow to my ribs. They were very sore; I thought I had broken one and gave up trying to hoist the baby storm jib. I retired below to a cup of tea. This storm lasted 24 hours but by the next night it was calm again with the sails flapping in the heavy sea still running.

The damage to me was a cut hand getting the jib down, and some very sore ribs. The dodgers were badly torn and the deck fitting carrying the radio aerial was broken. I hoped it would not affect transmission. I lashed it down with some Bostick and wrapped it round with insulating tape. It proved effective.

By March 13th we were down to 50° S. and it was cold and rough – blowing a gale, misty rain and spray with visibility no more than half a mile. I was running under spitfire staysail, dozing on my bunk. I would hear a hissing and would brace myself against the shock as a sea would hit us and break all over. My ribs were sore and it hurt me to breathe deeply or to pull on the halyards. But we were heading eastward, always fighting to get eastward. Until now I had not worried too much if we went a little off course to north or south, but as we got nearer to the South American coast more care had to be taken. It was a long way yet to go though. I ate the last of the tomatoes, given to be at Bluff, with a tin of salmon and cold potatoes. Those tomatoes were some of the best I ever tasted and I didn't waste one, after five weeks.

March 19th saw us 52° 20 'S., 98° 40' W. and about 1,000 miles from the Horn. The sun crossed the Equator to north, so it was autumn down here, later than I wanted to be. We were nearly becalmed. The clew of the storm jib showed signs of giving out. I bound it round with tarred twine. I then topped the batteries up with distilled water; they wanted very little. The compass light gave out and I fiddled about with it but I failed to get it to work. I would have to use a torch to see the reading in the dark. I started the main engine to charge the batteries. The self-starter wouldn't work so I had to swing it by hand. That did my sore ribs no good. I traced all the starter connections, but found them all in order. After easing the starter button out of the panel, however, it worked, so there must have been a short somewhere, and moving the button freed it.

I had the last of the eggs from Melbourne; they had been marvellous – not one bad and only one cracked in nine and a half weeks.

*　　*　　*

On March 22nd we were down to 54° S. and were almost

becalmed in the light easterly. I sewed up a seam in the genoa staysail and finished on the foredeck with my hands so cold I could hardly feel the needle. A hot drink was the order after that. Then it blew up from the south-west and by the 24th it was blowing a strong gale. During the night the log line, apparently lifted by the sea, had become entangled in the self-steering gear and had chafed through, losing the rotator. I had a spare one and rigged that. It was cold, with thick mist and visibility down to a quarter of a mile. My dead reckoning position put me as about 700 miles from the Horn. I hoped the weather would clear, so that I could fix my position approaching this rocky coast. The glass went up and down like a Yo-Yo as calms followed gales. I had got one shaky sun shot on the 25th which on my dead reckoning put me as about 150 miles south-west of Noir Island on March 28th. This was the nearest land off the south-west coast of Tierra del Fuego.

On this day there was another dreadful incident with the fumes from the little generator. I had run it to charge the batteries for my attempt to contact Puntas Arenas radio. For the second occasion the exhaust pipe became disconnected and when I went to stop the engine I was again overcome by fumes. Apparently, I had not learned my lesson the first time and it was great carelessness on my part. I regained consciousness, not knowing where I was, but slumped at the bottom of the companion steps, with my right leg doubled under me and I collapsed completely. After a struggle I got to my feet only to find I could not put any weight on that leg at all. My ankle just buckled up. I got my seaboots off and massaged my leg. It was not painful, so no bones were broken, or sprained. It was blowing a strong gale and I had taken all sail off before the incident, so I left her to look after herself and turned in. I kept massaging my ankle to get the blood flowing again. I had no feeling there at all.

The next morning dawned bright and clear. I managed to get on deck and set the staysail and storm jib, but I had to be very careful, with the movement of the yacht, not to throw my weight on that ankle which just doubled up at the slightest provocation. I managed to get a shaky sun shot, which put me as 55° 15′ S., 75° 05′ W.

I was heading to pass south of the Ildefonso group of islands and north of Diego Ramirez Islands. There are no lights in these groups and the gap between them is less than forty miles wide. That night, March 29th, I tried to contact Punta Arenas radio, but *Wave Chief* picked me up and came in offering to relay any messages. On my previous attempt at making contact with Punta Arenas I had heard the R.F.A. tanker *Wave Chief* calling them. At that time I had no idea of the identity of *Wave Chief* or of the fact that she was on the lookout for me.

The next morning, the 30th, I found my leg much better and stronger; the blood seemed to be circulating and I had some feeling there. By noon it was blowing up, but I managed to get a shaky sun shot, in heavy seas, violent motion and spray. It put me as 55° 45′ S., 71° 55′ W. I lowered the staysail and ran on under storm jib. It was rough going, with big seas coming at us on our quarter, swinging us round and laying us well down. By dusk I reckoned it at force 10 gusting perhaps 12 from north-west and I was forced to lower the jib and lay-to. I heard *Wave Chief* calling me on the radio telephone. I made contact and received a message from Dorothy and also the British Ambassador in Chile, wishing me well. I passed my position and the weather conditions to *Wave Chief*. She was getting much the same rough treatment.

This strong north-westerly gale continued all night until noon next day when it moderated enough for me to set the spitfire staysail and also to get a noon sight which put me as 56° 05′ S. I kept a sharp lookout for land. I spoke to *Wave Chief* who asked me my position and also if I could take a direction-finding bearing on a signal they would transmit. This I did and it put me due west of her.

Just at this time I heard a call from a plane, looking for me. It was the Independent Television News and *Sunday Mirror* reporters. They asked me my position and also to call them up should I hear them in the distance. Meanwhile the weather had moderated and I was able to set the storm jib and also the trysail. While I was doing this the sky brightened to the north and there on the horizon was the outline of jagged rocky mountains with a grey-blue haze over them. Also at this moment a plane appeared heading right for me. It circled several times and took photographs, I presumed. I heard them call *Wave Chief* and tell them I was due west of them. The plane then flew off to the north, heading back to her base. I asked them to identify the land to the north for me and I was told it was Waterman Island some forty miles off.

I was just about where I thought I was and between those two groups of islands that I wanted to miss. Well placed, in fact, for my approach to the Horn. I never did sight these islands. During the afternoon I sighted a tanker approaching from the east. It was *Wave Chief*, of course. She had been detailed to watch out for me rounding Cape Horn. She circled round and took up station astern of me and on my port quarter. It was a quiet night, with a light north-westerly wind and a moderate sea. *Wave Chief* kept station with me and we had frequent chats on the radio telephone. My leg was much better. At dawn on April 1st Cape Horn stood out majestically from this awful rocky coast, bearing north-east distant about twenty miles. I was running under working jib, staysail, and trysail. It was noon when I actually rounded the Horn, about 11 miles off. Another plane appeared and circled us. April 1st was the date Dorothy had forecast to the Press that I would round this southernmost tip of the South American continent, and it was certainly an April Fools' Day I shall never forget.

I stood and stared at that great hump of land. This was it. This was the moment I had dreamed about and planned for. I thought of all the others who had passed this way. Lone sailors as well as those in the great square-riggers. I was just another one, looking with awed respect at this most feared of capes.

I was lucky with the weather. A light north-westerly and a moderate sea drove me round comfortably, but it was cold. One could feel the awful desolation and cruel bleakness of the area. To be driven on to those rocks in an onshore gale would be certain destruction. I looked at *Wave Chief;* a big powerful ship she was, but she was insignificant against the background of mountainous rocks. Nice as it was to have her company, and grateful as I am to the Admiralty for their kind thoughts in sending her along, I doubt if she could have done much to help me if I had been overwhelmed in a gale such as I had a couple of days before.

I went below and made myself a hot drink of lemon, honey, and a stiff tot of whisky to toast my rounding of Cape Horn. I had been given a bottle of champagne for this purpose, but it was cold and I did not fancy a cold drink at the time. The hot drink went down well and warmed me up inside. As I sipped this a cold misty rain came down and a great black cloud descended over Cape Horn, blotting it out of my view. I altered course more to the north-east. Every mile now would take me north, to where the sun's rays had some warmth in them, away from this cruel Southern Ocean. At dusk we were about 15 miles east of Barnevelt Island to the east of Cape Horn. The wind backed to the south-west and I lowered the trysail as it was covering the jib. I passed the Le Maire Straits and coasted along about ten miles off Staten Island as dawn broke on April 2nd. The wind was south-west dead aft and I was running under boomed-out working jib and genoa staysail on the other side. As we passed the northern tip of Staten Island at about noon *Wave Chief* called me up on the radio and said she was now leaving me to return to her base in the Falkland Islands. She closed with me and the crew lined the rail to give me three cheers. I was greatly moved by this demonstration of goodwill and kindness. Then with three blasts of her siren she was away. I called up the captain on the radio and thanked him for escorting me and for his courtesy.

I4 *Variables of the South Atlantic*
For maps see pages 103 and 119

As *Wave Chief* drew ahead and disappeared into the mist I felt quite lonely. It made me realize what an empty ocean this is, and what a long haul I had ahead of me. A sailing distance of almost 8,500 miles, and I was making such slow progress.

That night, April 2nd, I spoke to Port Stanley radio in the Falklands and received a message from the *Sunday Mirror* and also from Dorothy which cheered me up. I was heading to pass east of the Falkland Islands, and had inquired from Port Stanley whether there was any ice danger. It was reassuring to hear that it was quite clear. As there is a current which runs up round the west coast I headed a little more east, to make sure of clearing Beauchene Island, which is a small island lying off the south-east coast of the Falklands.

The wind had gone to north-west and I was running under working jib, genoa staysail, and trysail, and by the early hours of the morning of the 4th the wind had freshened so that we were knocking up seven knots. I lowered the genoa staysail without loss of speed.

On looking round the gear soon after dawn I received a shattering blow.

I noticed the servo-box carrying the servo pendulum blade was badly distorted and twisting under the strain of the blade. I looked at it and watched it, a cold fear clutching at my heart. It looked certain to go in time. The point was how long would it last.

I visualized myself sitting and steering for hour after hour, day after day and week after week. Should I put into Port Stanley? I dismissed the idea for the time being and held my course. But I was worried. Morale was at a low ebb just then. I spoke to Port Stanley that night and also *Wave Chief*, back in harbour, and exchanged pleasantries.

The next day, April 5th, was calm and sunny with a light north-westerly breeze giving us about five knots, and that night I got a message of 'Good luck' from the people of the Falklands via Port Stanley radio. I was then about 120 miles due east.

I watched the self-steering gear constantly – especially when it blew up to gale force 8 the day after. I came to the conclusion that if I could fix a block of wood between the two walls of the housing this would stop the twist. I hunted through my lockers and found some odd offcuts of wood, and one piece seemed to fill the bill.

The next day, April 7th, it was flat calm, with bitterly cold, wet, dripping fog. I got the self-steering gear inboard and fitted the piece of wood between the two sides of the servo-box. After a little attention with a file it was just a nice tight fit. I drilled about four holes each side and screwed the sides on to it with brass screws. The sea was still lumpy and it was an awkward job, on a rolling deck. However, I did it, and reassembled the gear, hopefully. That night I spoke to Port Stanley for the last time.

A fitful breeze came up from the north-west, giving us just steerage way all night, but by noon next day it had increased to force 5 or 6. We slid along at a great rate under boomed-out jib and genoa staysail the other side, covering 156 miles during the next 24 hours. This was one of the best day's runs. It was hectic going and at times we were beating seven knots.

The wind dropped away, however, but after a calm it veered to the north and blew up to gale force. We were close hauled with solid water over the foredeck. It was April 11th and in the ten days since rounding the Horn we had covered 1,000 miles. But I felt we ought to have done better, somehow. Then I am never satisfied, am I? I noticed the self-steering gear was twisting again and found the screws had sheered off on one side. When the weather moderated I had the gear off again, drilled a hole right through, and bolted it together. It was difficult work, with a breast drill, in a seaway. One needs two pairs of hands. I broke one drill when *Lively Lady* gave a lurch, and I twisted the brace. However, I managed the task and felt more confident of success this time. I got the gear re-rigged just in time for a sudden squall that blew up from the west-south-west. I was forced to lower the trysail and the jib, as it continued to blow up to force 9. This continued for two days, varying between south-west and south. On the 12th April I wrote that night in my log:

A wild, rough night, the wind – bitterly cold – screams at us; the sea boils as the tops of the waves are blown off, and the hail hits the deck like bullets . . . The whole ship shudders and shakes as the wind roars through the rigging. Big seas lift our stern and swing us right round off course, and three times I had to scramble into oilskins to go on deck and put her back on course. In the end I was forced to lower the staysail and lay to.

My hands were so cold that I could hardly write up the log. That called for a hot drink, which was a great joy and warmed me up. We were being rolled about unmercifully, and one bad sea laid us so low that it broke the lamp glass of the little oil lamp swung in gimbals – the first time that had happened. However, I had a spare one. I find a small oil lamp very good in the cabin at night. It gives a low glow, as compared with the bright glare of an electric lamp, helping the eyes when called outside at night. Of course, one has to have a good light for chart work.

After another night of fierce winds, hailstorms, and big seas coming over the side decks and filling the cockpit, it eased. I was able to get under way with the staysail. It was still blowing gale force 8 and the big seas would get under our stern and swing me round. She would not come back without assistance on the tiller.

At dawn on April 15th it showed heavy grey cloud, and big, white-capped seas rolling up. However, by noon the sun broke through and I was able to get my first sun sight for four days. It put us up to 40° S., and I was hoping that it would be getting a little warmer. It continued bitterly cold, though, with the wind round to south-east. The albatrosses were still with us, also some smaller birds, grey on the back and white underneath. They had a quick wing action, diving and darting about like swallows. I saw several schools of small black whales. One lot crossed close astern, and I felt sure one of them would foul my log line. They did not touch it, though.

It was on April 17th that it really began to warm up, and I had to shed a woollen jersey. I began to feel we were making real progress northwards. It fell flat calm, though, and by noon on the 18th our week's run had been only 470 miles. Then the wind came in from the north-east and we were close-hauled heading east. I wanted to get east, but I also wanted to get north. The wind gradually increased until, after changing working jib for storm jib and reefing the mainsail, I was forced, on the evening of the 20th, to lower the jib and the mainsail. By midnight it was blowing up to force 10. My log reads:

> What a terrible, rough night; a couple of seas hit us hard, it threw everything across the cabin, turning the yacht round so that she was hove-to with the staysail. Worry was the wind vane, which turned itself round, so it must have slipped.

Lively Lady would run with a following sea quite true, but with a big cross sea coming up and catching her under her quarter it would swing her round. The self-steering gear would not always bring her back on course, as she would tend to come up into the wind. It was then that I had to assist the self-steering gear to get her back again. On this occasion the seas that hit us were so violent that they carried the yacht round bodily until she was hove-to on the other tack. The wind vane of the self-steering gear is designed to slip in the clamp which holds it under extreme pressure, such as was experienced in this case, to prevent damage.

I was forced to lower the staysail in the end. This carried on for 24 hours, when it moderated and backed to west.

We were now in the belt of variables: the fickle winds and weather which occur north of the westerlies of the Forties, until the south-east trade winds are met in the Twenties. There were some 1,200 miles of these ahead of us to get through. For example, on the 22nd, it was blowing from the south-west, and we were running under working jib and mizzen staysail.

Then it fell flat calm and I lowered the staysail. Next came a breeze from the north-east and I set the mainsail. This increased to gale force 8 in fierce rain squalls, and I had to reef the main-sail and lower the jib. I set the spitfire staysail and we were close-hauled. Heavy rain and a thunderstorm blew up, fol-lowed by a flat calm. My log reads on the 24th: 'What a day – blowing a gale one minute and then a flat calm. Just had to lower the mainsail in flat calm; but even as I write, it is blowing up.'

And it did indeed, blowing a strong gale from the south-west. I had to turn out in the dark hours of the night to lower the jib. The wind blowing force 10 fairly took my breath away, and solid sheets of water rolled over the yacht. This kept on until noon of April 25th. Our week's run had been under five hundred miles; very frustrating. 'Feel listless and tired,' I wrote in my log, 'everything too much trouble to do. Morale is low, due to our slow progress. A long way to go yet and it seems as though I shall be for ever doing it.'

I had to dismantle the steering gear again on the 27th, when I found the swivel pins of the steering servo pendulum blade loose. I had to take the distance piece the pin is bolted through right out to tighten it. While I was doing this the wind went from west to south, and then south-east, east and north-east. It was light and variable; whichever tack I went on was the wrong one. On April 29th I ate the last of the oranges from Melbourne. They had been very good; a few had gone bad, but then they would have been last year's crop, so they were pretty good really.

My position on April 30th was 25° S. and 29 °W. I wanted to get farther east to clear Martin Vas Islands some 600 miles off the South American coast and to pick up the south-east trade winds.

It was getting quite warm and I shed my woollen sweaters and long trousers for shorts. The wind continued light from the north-east, and we were close hauled the whole time, heading into a choppy sea, with a wet foredeck. The working jib started to pull out at the tack, so I changed that for the new spare one I had got Mr Lucas to make for me before the start. The nights were getting shorter and the night of May 1st was brilliant with stars and the Milky Way brighter than I have ever seen it. A crescent moon shone out, and it was warm and comfortable.

It was not comfortable for long, though, as at noon the next day it blew up. I was just about to dish up a steak and kidney pudding, potatoes, and carrots, when the wind suddenly backed to the south-west, and a fierce squall came in with heavy rain. I had to lower the mainsail and working jib in a hurry, and we were soon running under spitfire staysail, with a gale force 8 blowing and seas building up. By the next morning we were becalmed and I sat on the foredeck and sewed up a seam in the genoa staysail. As I finished the wind came in from south and backed to south-south-east, and then south-east, very light. I set the working jib, and the mizzen staysail.

On May 5th my position was 20° S., 26° W., and I had passed Martin Vas Islands leaving them some one hundred and fifty miles to the westward. I still wanted to make more ground to the east, however, as a strong current sets to the west, increasing as we get farther north, to up to two knots per hour. I picked up my first flying fish off the deck since rounding the Horn, though I had seen them about. It made a nice change for breakfast, fried in butter.

The winds continued very light to flat calm. I hoped we had picked up the trade wind when it came in from the south-east. But it fell away to calm, then a light breeze from the north-east, followed by another calm and a wisp of air from the south. We were running under boomed-out genoa and mizzen staysail when we touched our outward-bound track on May 7th, 1968, having completed the circumnavigation of the world since September 6th, 1967: eight months.

The weather was hot and humid, the sea calm and placid. I saw very little wild life. The odd few seabirds seemed as lazy as the wind and sea, and flew round or sat on the sea, looking thoroughly bored with life. An occasional dolphin's fin would break the surface, but even they were not in their usual playful mood. I observed flying fish taking off and gliding effortlessly through the air for surprisingly long distances, in their bid to escape their pursuers.

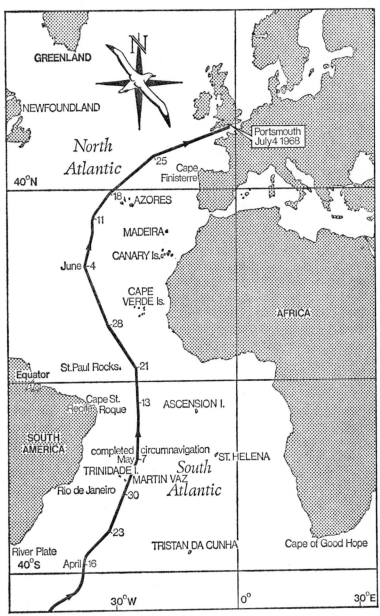

7 South Atlantic to Portsmouth

May 8th found us back close hauled again as the wind went to north-east, then to east, and increased to force 4 to 5. We were up to 15°S. and it was time we picked up the trades. I hoped this was the beginning. It was quite a rough night, with a boisterous sea, plenty of water on deck as *Lively Lady* was laid well over at times. Noon-to-noon run was 141 miles, the best for a long time. We were well off the main shipping lanes, from the River Plate and Rio de Janeiro to the English Channel, but crossed the routes from the South American ports to Cape Town. I kept a lookout for signs of ships, but did not see any.

Our next day's run was 147 miles, and 150 the day after. At last we really were in the south-east trades, and was I pleased! There was more north than south about those easterlies though. It was boisterous going, with quite a violent motion at times. I kept a watchful eye on the self-steering gear, as it was under considerable strain in these conditions; however, my repairs seemed to have stiffened it up quite a lot, and I felt reasonably confident that it would see me through. I ate the last of the apples. They had lasted well, but had begun to go soft.

They were bright, sunny days and clear, starlit nights as we sped along across the trades. We logged over nine hundred miles in seven days to May 13th, in spite of being becalmed in heavy rain for about three hours on the 12th. I took the opportunity of having a cold shower during this period. I stood out on deck and soaped myself down in the heavy rain. It was very refreshing.

We were now about 4°30′ S. and 24°20′ W., and roughly 650 miles east of Cape St Roque, the north-eastern point of South America. The wind went very light and fluky to the north-east, and our speed dropped considerably as we headed into the seas on our starboard bow. We logged up our 10,000 miles from Melbourne, in 121 days, including five days spent in Bluff; an average speed of 86 miles a day in sailing time. This was most disappointing, but the head winds I had encountered on leaving Melbourne and again on leaving Bluff, when I had those strong easterlies for three weeks, knocked back the average. Again the slow progress across those variables in the South Atlantic also took its toll before I picked up the trade winds.

It looked as though we had lost the trades and were in the Doldrums when, on the 14th, we were becalmed in latitude 3° S. In heavy rain I took another cold shower. I also took the opportunity of servicing the main engine. I turned the engine regularly every day to keep it free, and ran it about once a week to keep it dry, and to keep the batteries up. The carburettor was flooding, so I had that off and found the needle weights seized up on their pivot. I freed them, had the plugs out and cleaned them, and filled the engine up with oil. I also cleaned the fuel pipe filters. There is always something to do.

The wind came fitfully from all directions of the compass and I was kept busy trimming sails day after day. Somehow we ghosted along north, towards the Equator.

15 *The Last Lap*
For map see page 119

WE crossed the Equator at about 0100 on May 19th. I invited Father Neptune on board to cut my hair, but he did not make a very good job of it. However, I toasted him in a large glass of sherry, and Algy joined in as well, making a few rude comments about some people's efforts at hair-cutting. I silenced him with the threat that I would invite the barber to shave his whiskers off.

The dawn came clear and bright, and a crowd of dolphins played round the yacht, having great sport.

We had crossed the Line on longitude 26°W., farther west than I had wanted to, but the northerly wind and the strong current running up to two knots had carried us westward. I was well clear of St Paul Rocks, though, being about 180 miles east of them. These are a dangerous group of rocks rising out of the sea just north of the Equator, and feared by the old sailing-ship men in their run across the Equator, whether heading north or south.

After a fresh westerly on the 20th, in which we made a run of 130 miles to noon on the 21st, it was calm again. It was tacking all the time as the wind varied from north-west to north-east. The weather was just as varied, with low cloud and rain. I was five days without a sun sight. I wrote in my log: 'The days go on and mount up – we seem to make such slow progress. It seems like a life sentence of sailing, tack and tack, calms and light head winds.'

At last on May 24th, in latitude 6°24' N. and longitude 26°50'W., the wind came in steady from the north-east, and I wondered if we had picked up the north-east trades. The sky was overcast with low cloud, and the wind was much cooler. We were close hauled heading about north-west, and, when on the 25th the wind blew more from the east, it allowed us to head 10° freer and our speed increased. The sea was boisterous and the motion violent as we sped along across the north-east trade winds with the rail under at times. But it was exhilarating sailing.

On May 31st we had run nine hundred miles across the trades in seven days, and to celebrate I picked a flying fish off the deck for breakfast. This was only the second one since leaving Australia. It was indeed an empty ocean down there: no birds, no sign of life of any sort for days on end.

At times it was rough going. The wind seemed to freshen at night, and we bashed along, making it very uncomfortable below. It was difficult to write up the log at times. We had passed the latitude of the Cape Verde Islands – but some five hundred miles westward of them.

I was still eating well, and had plenty of water in hand. After my usual cup of tea at dawn, laced with whisky, I continued my routine of Shredded Wheat or porridge for breakfast (plenty of sterilized milk from Australia), followed by fried onions or a tin of sausages with baked beans, Ryvita or Energen Crispbread biscuits. During the morning I would have a slice of fruit cake and coffee or cocoa. At lunch it was a stew of tinned stewed steak with baked beans, tomatoes, carrots and onions added. Other variations were: tinned steak and kidney pudding, corned beef, salmon, herrings in tomato sauce – all with vegetables: tinned potatoes, mashed potatoes, baked beans or something of that kind. A cup of tea was taken in the afternoon, followed by supper of biscuits, cheese, butter, and a raw onion, biscuits and honey, with a slice of cake to finish with. I had plenty of tinned fruit cake, which opens fresh and is good food value. During the night – or at any other time I felt the need of it – I had my favourite hot drink of lemon juice and honey with hot water and topped up with whisky. I also had some sweet biscuits, such as shortbread in tins. So you can see that I ate well. I also had a can of beer at lunchtime, but when it was very cold I had my hot drink.

We sped along north-westward, and on June 2nd I caught up with the sun on its northward journey. At noon that day I followed it right round the horizon when trying to take a position sight. I had to give it up in the end. I got a sight later and fixed my position with a cross bearing.

The next day the wind fell very light and I logged: 'It looks as though we are running out of the trades.' There was a flying fish for breakfast, though.

I noticed the log line had a row of barnacles sticking out of it, and I hauled it in and cleared it. The wind was very light to calm and we ghosted along in a general north-westerly direction, hardly making steerage way at times. It was lovely weather otherwise: clear sky, calm sea and at night the stars shone out clear and bright. I busied myself sorting out stores below. I topped up the batteries with distilled water and it was surprising how little they wanted. I emptied a jerry can of water into the small water tank under the deck. I was using this for drinking purposes, as the remains of the water in the main tanks had got contaminated by sea water. As I mentioned before, when seas swept aboard the sea water got in through the overflow pipe which came out on deck in a small copper pipe. I had closed the end of the pipe, leaving a very small opening for air, but the sea still got in and it is surprising what a small quantity of salt water is required to contaminate the whole tank.

By June 6th we were level with the Canary Islands, though about a thousand miles west of them. Great patches of weed would float by. It fouled up the log rotator, and I was for ever hauling it in to clear it. It hung round the servo-blade of the self-steering gear, and I had to push it off with the boathook.

On June 8th, great excitement. I saw my first ship since leaving *Wave Chief* at Cape Horn. The M.V. *Sunseahorse* turned and answered my signal of 'M.I.K.', the coded signal asking, 'Please report me to Lloyd's, London'. My position was 30°N. 37°W., about six hundred miles S.S.W. of the Azores. The wind went round to south-west, and I boomed out the big genoa and the genoa staysail. It was very light, but we made fair progress through the night. I think it was the first free wind we had had in the 6,000 odd miles since leaving Staten Island, but then it went to north-west, and we were back close hauled.

We had very light wind, calm sea and a clear sky, with bright, moonlit nights until the 12th, when it blew up strongly from the north-east. Black, low clouds scudded by and there was a whine in the rigging. It was much cooler and I had a woollen jersey on. Quite a sea got up as we headed into it. The wind varied in strength and direction, backing from north-east to north-north-west; right on our nose the whole time. It was a succession of tacks, heading into a lumpy sea, with low cloud and rain and pitch-black nights. Desperately slow progress. On the night of the 14th it blew up to force 7 and I had to turn out in heavy rain to reef the mainsail. By dawn we were becalmed, and lay for a whole 24 hours, wallowing in a lumpy sea, with misty rain. However, by May 17th we were on the same latitude as Flores Island, the most western of the Azores Islands, but we were some 75 miles west. I tried to contact them on the radio telephone, but without success. I could start turning north-east now on a great circle track for the English Channel but the next day the wind came north-east – dead ahead. Whichever tack we went on was the wrong one. But we were heading too far east really.

It was terribly slow going. The sea was like glass. How we made any progress at times I don't know. We just ghosted along. The breeze, or light air I should say – one could hardly call it a breeze – backed to west, and I turned north, hoping to find some wind. We were now about 100 miles north-north-west of Flores Island.

Not much life was about. One day I passed through a whole flotilla of Portuguese 'men-o'-war', their purple tinted sails set hopefully, though where they were bound for I do not know. I don't suppose they knew themselves. Then a school of dolphins played around for a short while. They always seem so happy and full of fun. I observed a couple of white tropic-birds circling overhead. They were slim with long white tails. Very graceful I thought them.

At last on June 21st the wind backed to the south-west and freshened. I set the big genoa on the port side boom and the genoa staysail on its small boom to starboard. Then I lowered the mainsail and we were running free.

We picked up speed as the wind freshened, touching seven

knots at times. Heading north-east on a great circle track for the English Channel, I felt we were at last within striking distance of home. Portsmouth lay some 1,400 miles ahead. I began to work out an estimated time of arrival, but gave it up at once, realizing it is fatal to do that sort of thing in a sailing boat at sea. But I began to get excited.

However, we were getting on well. As the wind increased so did the sea of course. It was boisterous, giving us a rough ride. The rolling was dreadful. *Lively Lady* would gradually build up the roll from side to side until she dipped her side decks under, then stop for a while and start it all over again. The seas, rolling up on the quarter, would lift her stern and swing her round. It was a pitch black night as we fairly romped along. The bow wave, the log line trailing out astern, and the white-capped waves were all a brilliant phosphorescence. Dawn showed a grey scene with misty rain and poor visibility. The wind had increased to force 5 or 6. I had hung on to the big genoa all night, but now the wind had veered to the north-west and I could no longer hold my course with a boomed-out sail. I stowed the genoa and set the working jib and the mainsail. By noon we had had a fair day's run of 147 miles, followed by one of 142 miles the next day. With the wind back to south-west we were running once again with the big genoa. I sewed up a seam in the working jib on the foredeck.

The night of June 23rd was pitch black and dirty with rain. It blew up to force 6 and I had to stow the genoa and set the storm jib for the night. The next morning it eased and I went to boom out the working jib. I found the boom had broken away from the slide on the mast. It had gone at the welded joint, just as the other one had. It seems that there was a weakness in the fitting here. This was a nuisance as I couldn't boom out my headsails at all now. However, with the wind to west I set the working jib and mizzen staysail. Another good day's run of 142 miles.

A fine night followed, but with a boisterous sea. It was about four o'clock in the morning when I heard a loud bang. I rushed to the cockpit to find the mizzen staysail trailing over the side. The halyard had parted at the bight where it goes over the sheave in the mast. I hauled the sail in and bagged it up.

The next few days were uneventful. The wind varied from south to north-west and from calm to near gale force. June 28th was a dirty day, blowing gale force 8 from the north-west for a few hours. Big seas built up and broke over into the cockpit, but by nightfall it was flat calm. Listening to the radio I was surprised to hear that planes had been looking for me in the Western Approaches. Why so soon I did not know, at the time.

By the 29th I was within about 160 miles of Land's End radio, and decided to try to contact them. I called them up, and Valencia radio came in straight away. I sent a message to Dorothy as well as to the *Mirror*. The next morning was grey and overcast after a dirty night. I was keeping a sharp lookout for land. With no sun sight for two days, I was on my dead reckoning position which could be out in these calms.

I called Land's End radio, and gave my assumed position. Just after that HMS *Letterston* called me up and I gave them my position. They asked me if I had heard aircraft overhead, but I hadn't. They then asked if I could take a direction-finding bearing on a signal they would transmit. I agreed – but before I could do so an aircraft flew low overhead, and dropped a smoke signal in the sea near me. I immediately called up *Letterston* and told them. We were nearly becalmed and there was thick mist and low cloud, but the aircraft had spotted me and directed *Letterston* to me. The fog lifted a little and we saw each other and made close contact before down it came again as thick as ever and *Letterston* was blowing her siren.

16 *Home Again*

AFTER meeting up with H.M.S. *Letterston* on Sunday, June 30th I felt I was nearly home again. I was emotionally moved when her crew lined the deck and gave me three cheers.

What had I done to deserve this? Sailed round the world?

Yes, but I had achieved the voyage because it was a lifelong ambition. I did it to please myself so I felt I scarcely deserved this welcome, but it was a nice warm glow that went through my heart all the same. I looked round at *Lively Lady* and felt that she was sharing my joy.

My mind went back to all we had been through together. The great gales and the calms, the night we nearly lost her mast in the middle of the Southern Indian Ocean; our arrival at Melbourne and our departure; our forced stop at Bluff, and the great moment when we rounded Cape Horn after a fearful storm the day before. Then followed the frustrations of the calms in the variables of the South Atlantic, the doldrums on the Equator, and the horse latitudes in the North Atlantic. At last came the good runs we made across the trade winds when hopes ran high of a fast passage.

All these thoughts went through my mind as the cheers echoed and re-echoed in my ears. I was frankly disappointed in my time. I had hoped to be faster. Was it I who had failed *Lively Lady*? True, she is not a fast boat, and we had had bad luck with the winds. I've always maintained that one needs a deal of luck in sailing to get the right winds at the right time. We did not get that luck. From the start we had head winds and calms for three weeks; then long spells of head winds off the Cape of Good Hope and again round Tasmania and off New Zealand on the homeward run when, by all the weather charts, we should have had favourable winds. What of my handling of my yacht? Had I failed her? I suspected I had at times when I failed to increase sail quickly enough after a gale. But then at other times I carried sail too long and overpressed her. All these thoughts raced through my mind as I watched from the cockpit of *Lively Lady*, lying becalmed in the mist.

I called down to Algy, who was still in his bunk, that we now had a Royal Navy escort and that he should feel very proud. His reply was that he thought it time we 'spliced the mainbrace'.

Soon after this H.M.S. *Letterston* lowered an inflatable rubber boat, called a Gemini, with outboard engine, which came over to me bringing my friend, Sir David Mackworth. He came on board bringing a bottle of champagne. I was, of course, delighted to see him and we quickly popped the cork and sampled the champagne, which tasted good to me. He told me something of the plans that were laid on for my escort up the English Channel and my reception at Southsea. It was wonderful to hear of all the work that had gone into these preparations. He left me a portable radio transmitter and receiver with which I could talk to my escort and they could pass information to me. This was to prove invaluable during the run up Channel.

We were just south of Bishop Rock in the Isles of Scilly and the north-going set of the tide was taking us on to them. The visibility was so poor that nothing could be seen. I was relieved, therefore, when a light breeze from the north-east came in and I was able to tack south and out of danger. By dusk visibility improved and Bishop Rock light blinked out due north.

Wind was variable during the night and I was constantly trimming sails and adjusting the steering gear.

The second escort H.M.S. *Laleston*, took up station about two miles ahead on the starboard beam, to keep off oncoming shipping. This she did most effectively all the way up Channel until the Nab Tower. By this time three press boats were with us and they came alongside and we exchanged greetings.

At dawn on July 1st a fierce squall descended on us followed by a calm, but by the forenoon it was blowing up to force 7 and I lowered the mainsail. The Lord Mayor of Portsmouth, Councillor Emery-Wallis, called me up on the radio telephone, and told me of the welcome they were planning for me. It was good to hear him and it made me feel very humble.

As the weather eased the reporters and photographers from the ITN and the *Mirror* were transferred on board by the Royal Navy in the Gemini, for recorded interviews and pictures. I suppose they were on board for a couple of hours, after which I re-set the mainsail and continued on course.

That evening we had a terrific thunderstorm. It was of tropical intensity, with brilliant lightning. A fierce hailstorm blew up with stones as big as peas hitting the deck like bullets. The yacht was laid down and I was forced to lower the mainsail. We were now about twenty miles south-east of the Lizard and south of the main shipping lanes.

The morning of July 2nd dawned grey and overcast with a rough sea. But David, undaunted, came over to see me in the Gemini, bringing with him a splendid gift of a tankard of draught beer, from the Chief Petty Officers' Mess. I duly drank their health. We conferred about a possible estimated time of arrival. In my talk with the Lord Mayor of Portsmouth I had mentioned that I had expected to get in by Wednesday or Thursday, but Thursday to be safe. In view of the arrangements that had been made and the number of people involved and the interest shown, it would be better if we delayed our arrival until Thursday morning. We might have made it by Wednesday afternoon, but the uncertainty would cause waiting about by officials and public, who had all been so kind as to want to meet me. I therefore continued under easy sail.

However, in spite of another violent thunderstorm, with fierce hailstones, we made good progress. By mid-afternoon we decided to heave-to for a few hours as we were getting on too fast. What an anticlimax this was. After fighting for days and weeks to get east, deliberately to slow down. Oh well!

We had arranged to meet up with the Southern ITV launch *Southerner* the next day at a point some 25 miles south-west of St Catherine Point, I O.W. Our position about 25 miles south

23. Alec Rose's own picture of the Southern Ocean.

24. *Lively Lady* seen rounding the Horn from R.F.A. *Wave Chief*.

25. The Royal Navy, in which the author served for many years in war, escorts *Lively Lady* up the English Channel
and (below) 26. his local newspaper, *The Evening News*, Portsmouth, is passed aboard a she approaches home.

7. A wave for friends as
8. (below) the armada
gains numbers off Bembridge, Isle of Wight.

29. Part of the great fleet of small craft, afloat and in the air, that escorted *Lively Lad* into Spithead.

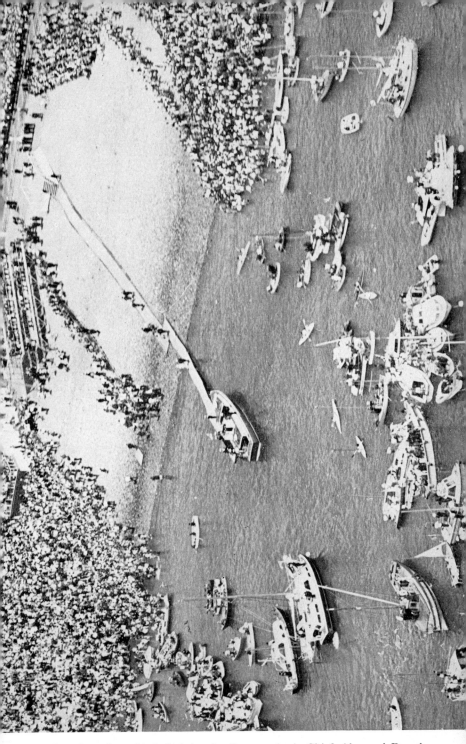

30. Brought ashore from *Lively Lady* by the Commander-in-Chief, Alec and Dorothy Rose walk up the beach at Southsea.

31. The citizens of Portsmouth mass to honour their new Freeman.

32. Civic reception at the Guildhall Portsmouth.

33. Cheers from the neighbours as Alec and Dorothy Rose appear on their balcony at home.

34. Rt. Hon. Horace King, Speaker of the House of Commons, receives Sir Alec Rose after he had been knighted by the Queen.

35. The author steps ashore to complete his solo voyage round the world.

of Portland Bill put us within a few hours' sail of this rendezvous.

As I was about to heave-to I saw a large vessel approaching on our port bow. Though close, I thought she would pass safely to port. Both *Letterston*, who was on my port beam, and I held our course. Suddenly the vessel altered course to port and proceeded to cross our bows. *Letterston* sounded her siren and I was forced to alter course to port to prevent a collision. She passed only about a cable's length across my bows, at about 15 knots. As her huge bulk passed me, I reckoned she was about 20,000 tons, and I read her name, *World Seafarer* of Monrovia.

Also about this time two French fishing trawlers came up and became a little too inquisitive. I observed Sir David put off from *Letterston* and board one of the trawlers. He told me later that after explaining what was going on he had suggested that as M. Jan de Kat had been rescued by the R.A.F. in the Atlantic, General de Gaulle would undoubtedly send the crew of any trawler menacing *Lively Lady* to the guillotine, on their return to harbour. This seemed to have the desired effect.

I hove-to and took the opportunity of having a rest on my bunk, *Letterston* offering to rouse me when it was time to make sail.

The wind died away to light at sunset and we got under way. It was a quiet night with a half moon and we made good progress. The morning of July 3rd dawned bright and sunny and by eight o'clock we had made such good progress that we had to heave-to again to await the arrival of the television launch. This had been delayed for technical reasons and was not due until 4 P.M. David came over and we discussed final details of the arrival plans over a glass of beer. It was all very exciting.

At about 4.30 P.M. the launch *Southerner* arrived and Bill Richardson and Barry Westwood, together with the necessary technicians, were transferred over by Gemini. The interview was put out 'live' and I understand was successful. I got under way with a single headsail with the object of reaching the Nab Tower at the entrance to the Solent at 8 A.M. the next day. However, by 9 P.M. that night the wind had fallen so light that we deemed it advisable to put up all plain sail. It was a quiet night and we glided along with only a very light breeze,

R—K

and July 4th dawned bright and clear, with a calm sea. Three fast patrol boats joined the escort and at 8 A.M. we were coming up to the Nab Tower. Then H.M.S. *Whitby* steamed by passing right through the fleet and the crew lined the rails and cheered. I dipped my Blue Ensign to her as she passed.

Already there were quite a lot of small craft out to meet me. David was ferried over to help me in the run in. The wind dropped light and *Lively Lady* was so blanketed by the mass of craft around her that we deemed it wise to run the engine just ticking over to keep us on course and to give us control of the yacht.

It was impossible to pick up the navigational buoys from among the crowd of craft around us and we had to rely on *Letterston* to guide us by the Bembridge Ledge to which we got rather close. By this time the craft around us was estimated at over four hundred ranging from small open canoes to large paddle-wheel steamers. As we approached Spithead more craft were observed anchored there. It was a most moving experience. The thing that impressed me was the good manners and seamanship of all the craft around. No unpleasant incidents were observed and everyone was friendly and happy.

The Commander-in-Chief's barge appeared with Admiral Sir John Frewen and Lady Frewen, bringing my wife Dorothy out to meet me. They preceded me in and tied up to a buoy opposite the Royal Albert Yacht Club. It was from here that I had left nearly a year before, when they gave me the starting gun. Now, as I crossed the finishing line and dipped my ensign to the Commodore's flag, I received the finishing gun.

I rounded up and, with David's help, stowed the sails and went alongside the C-in-C's launch, where Dorothy came on board to greet me. A great crowd estimated at over 250,000 let up tremendous cheers from ashore. Ships' sirens and yachts' hooters sounded and rockets were let up from the pier. Algy came on deck to wave, with me, in response. He became so excited and big-headed I was afraid he would fall overboard and I was glad to get him back safely below. Here I put a tie on and a reefer jacket to go ashore with Dorothy to meet the Lord Mayor who was waiting to greet me. We were then driven to the Guildhall where a civic reception was held, followed by a Press conference. Another crowd awaited me at my home when we arrived after a most wonderful and happy day. I was tired but a feeling of content passed over me. *Lively Lady* was taken care of by the Royal Navy, who put her into safe hands in the dockyard.

So ended the 'Great Adventure' in my life, but the memory of it will always be with me.

PART II

17 *When Alec's Away*
DOROTHY ROSE

W HAT people ask me all the time is whether I worry
and if not why not?

I am not a psychologist, but the whole thing seems
to me quite natural. Men are adventurous by nature, always
wanting to do something, though some never get a chance. It
was the same with Alec. His ambition was to sail round the
world single handed. He could not do it until his family had
grown up and even then he could not start until he had saved
enough, because even a small yacht like *Lively Lady* costs quite
a bit and preparations for a long voyage are expensive – at
least they were to us especially after all the setbacks.

My part was to share in the adventure and to back him up.
It was no use for me to worry, and I knew that if I did it would
spoil Alec's adventure and make him worry too.

Of course there were times when I was anxious. When a
message was received from Madeira that *Lively Lady* was in
distress I did worry until I heard that he was all right and that
the report was a mistake. Then again I had two restless nights
when *Lively Lady* was approaching the Foveau Strait in New
Zealand, as I knew if *Lively Lady* was in trouble Alec would be
worried about her, but once he got there I didn't worry any
more. You see, I had complete confidence in Alec as a seaman
and in *Lively Lady* as a good seaboat.

Another thing was that I had the shop to run. That kept me
busy. It might have been a different story if I had nothing to
do. You see I like our shop; I like seeing to things and I like
people. So I get a bit of everything.

Alec and I were both together in our thoughts about this adventure but our lives were a complete contrast. Take Alec's log, which I have only recently seen, of a Monday morning in the ocean:

'No let up at all – wind and sea terrible. The whole ship shudders and shakes with the wind. Dress up in oilskins to go on deck and see if I can get the baby jib up. Here goes!'

On that morning, like all other weekday mornings I would have got up at about 7 A.M. Then I would open the post – mostly people writing about Alec. At eight o'clock I open the shop to let my driver in. Then Monday is a cleaning-up day – all the fruit stacked and redone, and the frozen food cabinets all defrosted. Then the local growers who grow their own produce call and later on the Portsmouth market people ring up to tell me what they have, and what has come in on the lorries.

In the afternoon, Sheila, the manageress, goes down to market to see what they have and orders what we need. Sheila was a wonderful help. She's a sort of second me. If I am not there she's equal to me, in fact she is better than I am. She makes two and two come to four. She has left now to run her own shop, but she stood by me while Alec was away.

In the evenings I did the accounts and answered the letters. This was quite a job but it hadn't been too bad until Alec got to Australia. Up to then he had not been much in the news, as a lot of people thought he would never make it. But after he got to Melbourne the pressure began to build up because I got more and more letters and more and more people to see me. I answered every letter where an address was given. I had no typewriter and no secretary, so I always answered in my own handwriting. That usually kept me up to midnight and sometimes to one or two in the morning, but people were so kind that I had to answer them, even when they wrote to commiserate with me about what they imagined would be the loss of my husband.

Then there were all kinds of extra business matters to attend to. Alec and I are ordinary people and a world voyage was an expensive thing for us to finance. Apart from unexpected gifts from friends, Alec did the whole thing off his own bat and had no contracts of any kind until he got to Melbourne. It was then that people became interested; newspapers, television, and of course his book. Those were only the start and once round the Horn all kinds of things followed. One of the nicest was that Sam McGredy wanted to name a rose after him, and there were many other proposals as well.

I know nothing about contracts and such matters so I asked one or two friends to help me and they formed a Committee so that they could advise me. I hadn't a clue really and I would probably have put Alec in an awful lot of trouble if I had been left to attend to everything myself. Final decisions were left to me, but the Committee advised me very well and I could not have done without them.

People also started asking me to functions and to open bazaars and present this and that, which I always tried to do, so you can understand how busy I was. At the end of the day I just tumbled into bed and I had no time to worry.

Even if I had had the time Alec's friends would not have allowed me to. David (Sir David Mackworth) and Baba often dropped in to see me and so did other experienced sailing people. They knew there was a risk, but nothing to worry about, for there can be no estimated time of arrival for a small yacht in the ocean. It all depends on wind and weather. Even an extra month can go by and no harm done. It was the landsmen who worried most and the newspapers.

Besides this I had a lot of fun too and amusing incidents.

First of all I must tell you about Algy and the Leprechaun.

Algy belongs to David and Baba and they lent him to Alec as a sort of mascot for the Transatlantic Race in 1964. He seemed to have helped, as Alec was fourth in that race.

When Alec started in 1966 on the world trip Algy wanted to go too, and David and Baba gave him permission again to sign on for the voyage.

Just before Alec started two friends of David and Baba, the Irish actor Ray McAnally and his wife, came one evening. She did some music for him and recited some poetry, which she did beautifully. As a parting present they also gave him a leprechaun as another mascot. It's rather an ugly little thing really and I never did like it. The moment it arrived everything

seemed to go wrong. When Alec had to return to Plymouth for
the third time and the yacht fell over, and was so damaged that
he had to put the whole thing off for a year, there seemed
something about that was really hostile.

When I joined Alec at Plymouth this leprechaun was leering
at us in the cabin.

We sort of looked at one another and I said, 'You know, I
don't think that's bringing you any luck.'

'No,' he replied, 'I don't think it is. I don't think it gets on
with Algy.'

So we took it ashore at Mashford's. As we went through the
shipyard we noticed a furnace where they steam the timber
planks. We looked at one another.

Alec said 'Shall I?', and I replied 'Yes'.

So Alec put it in and cremated it.

After that nothing to speak of went wrong again. We might
be mistaken but I don't think Algy liked that leprechaun.

Long after that Mrs McAnally said to Baba 'If Alec thinks
that leprechaun isn't bringing him any luck, you'd better tell
him to dump it.'

Actually of course we had already dumped it but nothing
was said – though, mind you, it must have crossed her
mind.

That seemed a long time ago. There were plenty of amusing
things at the shop later on.

One day a schoolmaster came from Winchester and asked if
he could bring the children in. I said 'Yes' and they came into
the shop six at a time.

They were young boys aged about six to eight. I showed
them the chart with Alec's position marked on it. Then they
each bought a fourpenny apple and asked for an autograph.
We worked it between us and they did it six at a time. Sheila
took the fourpence and I signed the autographs.

The same afternoon a clergyman came into the shop. He
said he was a chaplain but he looked more like a pirate really.
He had got a patch over his eye and a peak cap at an angle. He
had composed some poetry about Alec and asked whether I
would like to hear it.

So I said 'Yes' and started to show him into the back room of the shop, but straight away he began reciting it then and there in front of all the customers. Of course they were highly delighted and he went through the two pieces of poetry and then asked if when I got in touch with Alec I would tell him. It was rather funny really to hear him reciting, this clergyman with a patch over his eye and all the customers getting poetry instead of fruit.

Besides the shop I had all sorts of interesting things to do. Attending to functions of all kinds, appearing on television, opening bazaars and so on. It was all rather an adventure as I was not used to it then.

But the great moments were when I had news of Alec. Then I would give a party to our friends and we would open a bottle of champagne as a celebration. Those were happy occasions.

The happiest of all was Alec's return. Both Alec and I were a little sad at having to wait so that he could arrive at his finishing line as planned in the daytime for his official welcome, but we did not want to disappoint people and it was worth while as it proved the event of our lives. He was given a naval escort and *Lively Lady* was accompanied by a fleet of yachts. Witnessing his great reception was my proudest moment.

Off Southsea beach, I was in the Admiral's barge which made fast to a buoy opposite the landing place. Alec, after rounding up, came back and brought *Lively Lady* alongside the barge. I then went aboard to talk to him in the cabin. It was the first time we had seen each other for nearly a year.

Somebody had put a bottle of champagne aboard in a special wooden box with ice packed round it. This was a great heartener because it was quite an ordeal for Alec to have to change over from being alone at sea for months with nobody to speak to and then suddenly to see over a quarter of a million people who had come to greet him. I was very excited about it too.

Anyway, we enjoyed the champagne and toasted each other in the cabin. Meanwhile, the huge crowd ashore were waiting and the loudspeaker was trying to keep them happy with remarks about the weather and what the announcer thought Alec was doing. He was not far wrong for, although he did not know that we were having a drink, Alec was in fact washing and changing into his shore clothes at the same time.

Then we were taken ashore by the Commander in Chief and given another tumultuous welcome, which was very moving. It was wonderful, and the weather was wonderful too. Everybody seemed so happy.

Then we went up on the dais and the Lord Mayor of Portsmouth gave us the official welcome home. Alec said a few words of thanks and we were driven off in the procession to the Civic Reception at the Guildhall and after that the Press Reception.

The first we heard about the knighthood was in the Guildhall on the day that Alec arrived. The Commander-in-Chief's Flag Lieutenant asked to see us in private. He said it was highly confidential and the Lord Mayor asked everybody if they would leave and he brought out a message from the Commander-in-Chief to say that Mr. Harold Wilson had put forward Alec's name to the Queen for a knighthood, but nothing was to be said until the Queen had consented to confer the knighthood. The next day the Flag Lieutenant came to the shop, but Alec was out as he had gone down to see the yacht. So the Flag Lieutenant came in and asked whether he could see me. I took him into the back room and he said 'Can I be the first to congratulate you, Lady Rose?'

Of course, I just burst into tears. On the day Alec arrived I hadn't. I had kept up very well – but when I heard that Alec had been knighted I just gave way. Then while I was weeping on the Flag Lieutenant's shoulder Alec came in and it was his turn to be congratulated.

Then a few minutes later the Lord Mayor arrived in state.

The announcement had been made over the wireless to the press and to the general public, and of course then it was bedlam. We had to shut the shop because everyone was crowding outside and we all went out on to the balcony. While he was away a Methuselah bottle of champagne had been given for Alec, to be kept for the day of his arrival. In the excitement I had forgotten all about this big bottle of champagne, but when everyone arrived to congratulate him I suddenly remembered and I said to David Mackworth,

'Could you pop down and get that big bottle of champagne? I think this is the time to use it.'

So he struggled out on the balcony with this Methuselah –

he had an awful job to hold it and to pour it. Anyway, he got it open. By then Osborne Road was so crowded that you couldn't get in or out, even with the help of police. Anyway, somebody in the crowd – I think it was a holiday-maker – got the cork and in the *Evening News* it said he was going to keep it as a memento. I was a little disappointed as I would have liked the manageress in the dress-shop right opposite to have had it, because she would have loved it. That was where my outfit for Alec's homecoming was designed and presented to me as a gift.

It was on the Friday that we heard about the knighthood so there was a rush to get ready to go to London on the following Wednesday for lunch with the Queen and the investiture, but the same manageress went round all their branches and they got me my outfit in time. You know I had to have something rather nice, but not showy, for the Palace and we were very pleased with it.

A friend put his car and chauffeur at our disposal and we got up there a little early so we waited along the Mall. We stopped the car and waited there. Alec got out to stretch his legs and of course everyone recognized him. They were passing by, then they would stop and turn round. You could imagine them saying 'That's Alec Rose'. We heard one say 'No'. I could recognize what they were saying. They argued between themselves and then they would come back and ask for his autograph.

Then when we got to the Palace there was another crowd outside the gates waiting for him.

It really was marvellous. First there were the sentries and the police outside. Then the going in and drawing up at the entrance and all the red carpet. The Master of the Household received us. We were taken upstairs and told what was going to happen, and what Alec had to do.

The investiture was simple. Only the Queen and Prince Philip and Alec and myself were there in a private room. After this honour the Queen gave Alec the photograph of the Royal Family, and she said when we went out the Royal Household wanted to cheer Alec, adding,

'If you go through the Grand Hall you will see the Royal Household lined up and at the end you will see a group of little

children, the Palace children. Two of them are mine and they
want to meet your husband.'

We went out and there were all the royal staff lined up each
side of the Grand Hall and they all cheered and clapped Alec.
Then when we got to the end there was this little group of
children – Princess Margaret's two and the Queen's two – and
Alec was introduced to them and then I was.

Then we went in to another room and had cocktails. The
Queen came in and Prince Philip talked to the other guests, to
whom we were introduced. Eight had been invited to meet for
the luncheon. Then we went into the lunch. When we came
out we had coffee and liqueurs, and met Prince Charles. He
had flown from Malta, intending to be at the luncheon, but
the plane was delayed by fog and he arrived too late, but
managed to come in for coffee. He had a lot to say to Alec
about the trip. We thought him very, very nice and very
engaging. They made a delightful family.

Afterwards we were told to stay on, because the photographers and the news camera people wanted to take us. In the
evening there was a reception at St James's for the Knight
Bachelors and as Alec was the newest one they all wanted to
meet him, so we went there. That was very nice.

Later on in the evening the Queen made an appearance and
we were brought forward to meet her again. Apparently she
had asked to speak to Alec again, and she said to me that it had
been quite a day for me, adding 'You have been to two palaces
in one day.'

We stayed on for a while before leaving. It had been quite a
wonderful day and then we stayed in London that evening.
The next day we opened the *Lively Lady* exhibition at the
Mirror. The bit that I liked was when Alec and I left the *Daily
Mirror* building and went to Greenwich to Trinity House for
the lunch. We had a police escort and the way was cleared for
us. We went over all the traffic lights, and everything had to
move out of the way for us. That was really fun. All these
traffic lights, whether they were red, green or yellow, were
cleared for us. Then when we got to Trinity House the Portsmouth Marine Band was there. They had found out that one
of our favourite songs was 'This is My Lovely Day' and when

we got to Trinity House the Marine Band played that for us. There was a whole crowd there, who shouted 'Well done, Alec.'

Altogether it was a wonderful week, the whole week was marvellous. Of course the welcome and meeting people were quite a strain on Alec who had been alone on the ocean for all those months, but he did not show it. Only one reporter noticed anything and wrote: 'All that day Alec had a far away smile. He was smiling but it was a far away smile.'

All I can really say for both of us is that the *happiest* day of our lives was when Alec arrived, and the most *memorable* day was when the Queen knighted him. So you see the whole thing has been a fairy story with a happy ending. I cannot say more.

18 'Lively Lady' At The Horn
RADIO OFFICER J. ROBERTS

BEFORE I butt into this story of *Lively Lady* I must introduce myself. I am a radio officer who qualified in 1964 at Wray Castle, Ambleside, and I have spent four years at sea in different ships visiting different parts of the world. It is an interesting job, because all communications with a ship pass through the radio officers, who thus are aware of much that is going on, next only to the Captain. So much so that one of the first requirements in a radio officer is the ability to keep his mouth shut. However, I have been given permission to give a short report and refer to the signals which led to the contact with Sir Alec Rose and his *Lively Lady* near Cape Horn; so here goes.

In March 1968 the Royal Fleet auxiliary tanker *Wave Chief*, in which I am second radio officer, was ordered to the Falkland Islands on routine duties and, whilst in the area, was asked to keep a radio watch for Sir Alec Rose. In this we were to relieve H.M.S. *Protector*, who was returning home from the Antarctic.

On March 21st R.F.A. *Wave Chief* was at anchor at Port William in the Falkland Isles, which is approximately 375 miles from Cape Horn, when instructions to proceed to Punta Arenas were received.

While there R.F.A. *Wave Chief* and the Punta Arenas radio station intercepted a weak transmission, believed to have originated from *Lively Lady* P.M. on March 26th.

Punta Arenas is the only town of any size in the Straits; it has a population of 46,000 and derives its livelihood from sheep (wool, skins, tallow, frozen canned mutton), and supplying

ships using the Straits and whalers from the Antarctic. Three
Russian survey ships left the harbour the day after our arrival.
There are two hotels, which provide about the only entertain-
ment available to strangers visiting these remote waters.

Here were the journalists waiting for news of *Lively Lady*. They
were in a very frustrated state of mind, for early reports of
sighting her appeared to have had no reliable foundation. On
March 20th the yacht had been reported at 56° 30′ S., 89° W.,
but this report proved wrong and the Chilean authorities had no
news. It was just a matter of sitting down and waiting.

It was not until March 26th that Santiago Press reported
that Third Naval Zone Chilean Navy contacted *Lively Lady* at
10 P.M. local time. Her position was estimated 200 miles west
of Cape Horn. This also proved false after checking with the
naval authorities.

Instructions were then received for R.F.A. *Wave Chief* to put
to sea and be on station twenty-four hours before *Lively Lady*
was due to round Cape Horn.

R.F.A. *Wave Chief* got under way on March 27th at 0630 local
time to start her search. She proceeded eastabout back through
the Magellan Straits bound for Cape Horn. At 0900 local time
on March 28th her position was fifty miles north east of the
Horn and the weather was worsening. There was still no contact with *Lively Lady*. The wind increased to gale force and by
nightfall it increased to Storm Force 10.

In my four years at sea I have encountered three storms. The
first was in a typhoon in June 1966 in the China Seas. I was
then a radio officer in R.F.A. *Fort Rosalie*, an 8,000-ton ammuni-
tion ship. The second was in a storm Force 10 off Algeria in
R.F.A. *Wave Chief* in December 1967. We were ordered to render
assistance, if required, to a Panamian ship which had suffered
a complete power failure, presumably due to boiler trouble,
which was eventually remedied, enabling the ship to reach
Oran. We made contact with her, but she signalled, 'All's well',
so we gave her the weather report and position and returned.

I was also in a storm which occurred in the Irish Sea in
February 1967. I was again in R.F.A. *Wave Chief*. She rolls and
pitches a lot, but is a very able ship in bad weather. You may
say the Irish Sea is relatively home waters and so it is, compared

with the distant oceans. But I thought it was the worst experience of the three. Perhaps this was because I had only just
returned from leave and had not recovered my sea legs.

Thus the storm when R.F.A. *Wave Chief* was approaching
Cape Horn was my fourth experience of really bad weather,
and it was something to remember. I usually enjoy a breather
on deck when it is really rough, but not at the Horn. It was
sunny but very cold and the wind seemed phenomenal. It came
in violent gusts which you could not face comfortably. All you
could do was to close your eyelids and hang on to something.
We were in a cloud caused by spindrift driven horizontally
by the screaming wind, with visibility down to less than a mile;
but these conditions moderated somewhat early on the morning
of the 29th.

Most of the time I was in the communications centre, taking
six hours on and six hours off with Peter Booth, the senior radio
officer. At 0900 R.F.A. *Wave Chief* was twenty miles east of Cape
Horn, and still no radio contact with *Lively Lady*. After we had
passed Cape Horn we were beginning to get a bit bored with
just sitting there and listening to nothing, which is what it
amounted to for, although a stream of talk in Spanish was
coming in all the time, it was meaningless to us.

Anyway, it was about ten o'clock at night, when I had been
on watch for four hours, when I thought I could hear a vague
little voice in the background of the other voices, and I thought
I heard the words 'Lively Lady' in it. I was not sure because,
for the past two or three days, when anyone said anything over
the radio that was fainter in the background, I immediately
jumped on and started trying to tune in because I hoped it was
Sir Alec. This exercise was becoming a bit frustrating after
being on constant watches for so long and not receiving a
clue.

However, this time it seemed different, and I shouted to the
navigator on watch who was working on the charts in the chart-
room next door. I said, 'I'm sure that was *Lively Lady*', and I
selected different aerials from the distribution box to the receiver and tuned in again. It was his call all right. 'Punta
Arenas. This is *Lively Lady*. Can you read me please. Over.'

My transmitter was on 'standby', so I switched it on and

called him, telling him we were the British tanker, R.F.A. *Wave Chief*, and asked if he could hear me.

He came back straight away and said he was picking us up loud and clear. I sent a message to the Captain. He came up and now our search really started.

First we asked Sir Alec his position, if he knew it after the bad weather he had come through, and whether he was all right. He came back and gave an approximate position and told us he had hurt his left leg while he was working on deck. He said he did not require any medical assistance as he thought his leg would heal up by itself.

An ominous crash, then a brief silence, was followed by a casual explanation that he had just been thrown across the cabin as a result of the yacht's violent motion, whilst talking on the R/T.

We chatted to him for about twenty to thirty minutes, and then told him that we would be keeping a constant listening watch on that particular frequency and there would be a person on watch twenty-four hours a day. Whenever he wanted he could come up on radio and there would be somebody ready for him. He must just please himself as to when he came up and when he trotted off. He told us he was about a hundred and ten to a hundred and twenty miles away, and replied about the weather, saying, 'Yes, it's blowing a storm here.' He then reported that he was sailing south-east at four knots.

For *Wave Chief* the worst of the weather was over. It had been blowing Storm Force 10 and was still gale or strong gale force, with huge seas running. No wonder we were surprised when Sir Alec calmly remarked that he was going below to write up his log but that he would give us a shout in the morning.

A yacht may be easy to locate in the English Channel in clear weather, but a boat of the size of *Lively Lady* is only a speck in the ocean when high seas are running to the west of Cape Horn and her position is only approximate. Most of the time she would be hidden in the seas. Making contact with other ships is a routine job in a Royal Fleet Auxiliary tanker, but with a yacht in the Southern Ocean it is another matter. *Lively Lady* was little bigger than a ship's lifeboat, except that she had a high mast. It was rather like looking for a needle in a haystack, but all of us, from the Captain downwards, were keenly interested in the exercise which was a change from ordinary routine.

The Captain's orders were to reduce speed from ten to twelve knots to eight knots and to make ten-mile sweeps at two-mile intervals across *Lively Lady*'s estimated course. From time to time *Lively Lady* gave fixes, but coming from a single-hander they were necessarily only very approximate. Sir Alec was always happy to cope with such irritating chores as the taking of D/F bearings under trying conditions to assist in making contact. We also received and passed on messages.

It was a complete day before we closed in and sighted her at 1600 on March 31st. At this time I was off watch and asleep. The news spread rapidly through the ship, and I pulled on some clothes and went on the bridge.

On the bridge there were five other officers. With the aid of binoculars I could just distinguish a tiny spar. This was the main mast of *Lively Lady* sticking up over the horizon. Then, when a big sea came and lifted his yacht up we could see, for a moment, the tiny blue speck of the hull. What surprised me was the small size of *Lively Lady*. The mast would often disappear for half its height in the troughs of the waves. She was reefed right down, and little of her trysail and storm jib could be seen.

R.F.A. *Wave Chief* closed *Lively Lady* at a distance of four cables. Sir Alec was in the cockpit, conspicuous in his yellow oilskins showing against the background of grey seas.

We kept station not less than five cables from the yacht at a fixed speed, opening up the distance when necessary by turning away, not by altering speed. *Lively Lady* assisted by burning a masthead light during the hours of darkness.

During the night we kept position by radar bearings. The navigator had to keep on the alert. *Lively Lady* jogged on as if we were not there. The weather was moderating all the time, and the visibility was invariably good whilst in company. Occasionally we had a talk with Sir Alec over the radio and he was extremely grateful to R.F.A. *Wave Chief* for relaying messages. He told us that his leg was getting better. *Lively Lady* had self-steering and much of his time seemed to be spent below.

He said he was doing a bit of writing, as he had a book to do.

Next morning, April 1st, the weather was even better and *Lively Lady* set her reefed mainsail and two running jibs. It calmed right off. But here I must add that a moderate swell was still running, for in the neighbourhood of the Horn the sea is rarely calm.

A signal was received from Ministry of Defence to *Wave Chief*: 'The Lady may be lively, but her chief escort is no laggard. Well done.' The reply was: 'Your kind message received and much appreciated by all on board. Fine weather now making life a bed of roses.'

Cape Horn was rounded at about midday on April 1st. Sir Alec was lucky for, although a swell was still running, it was calm compared with what he had been through in the approaches. It was sunny and we passed about ten miles south of the Cape. It was so clear that I was able to take photographs, one of which is reproduced in this book.

Our orders were to accompany *Lively Lady* and detach when satisfied she was in calm waters and clear of Cape Horn. The decision to part company at Staten Island was taken on board RFA *Wave Chief*. At the time of parting, the bosun came up to the bridge and asked if it was all right for the lads to say 'Cheerio' to Sir Alec. Permission was given to them to stop work, and they came trooping on deck and lined the rail on the side on which we were going to pass him.

RFA *Wave Chief* made a close run in. We closed in to one hundred feet when passing *Lively Lady*, and the lads gave him a cheer. Hearing this, Sir Alec came out on deck, waved goodbye to us and expressed obvious pleasure at this gesture. We then stood on straight on our course for the Falkland Islands.

Very, very slowly, *Lively Lady* receded in the distance. She became a dot and then, suddenly, there was nothing to be seen. Sir Alec Rose was alone again on the ocean.

The Cape Horn Region

19

MICHAEL MASON
(a former Commodore of the Royal
Ocean Racing Club)

T HE man who decides to round the Horn single-handed
is certainly a bold fellow. He is facing not the uncertainty
of the unknown but the well-known certainties of a
number of very desperate risks. Two things he can be quite
sure of: if just one important thing goes wrong he is a goner,
and if that happens all his troubles will be over very quickly.

My old friend Adlard Coles, who asked me to write this
account, was under the misapprehension that I had sailed a
small vessel round Cape Horn. This is not the case. My crazy
old boat would never have stood up to it. But for some months –
during the latter part of the summer of 1930 into the early
winter – my wife and I were cruising the region north and west
of the Horn, exploring the inland *esteros* (fiords, sea-lochs).

What had attracted us was an enormous map, in several
sheets, wherein the whole inland region of the Territorio de
Magallanes was described, from the summit-line of the Andes
to the open Pacific Ocean, as Unexplored, and many islands
were bordered by dotted lines. *Inesplorado*. Magic word! When
we got to Punta Arenas, in Magellan Strait (the only settlement
on the mainland there), we were encouraged by advice
(i) it couldn't be done; (ii) that we were off our chumps;
(iii) that the shores of those parts were so densely grown with
'virgin' forest that no human being could make his way inland
at all. We were 29 and 24. That is the right age for reckless
ventures, and success. And we did get to know the country and
the climate.

We were eventually frozen out of the inland *esteros* by the advance of winter. We came back by the open ocean, in part, and by the Smyth Channel (well surveyed, by HMS *Beagle* among others) and Magellan Strait. We spent only one night at anchor in the Tierra del Fuego archipelago, of which Cape Horn is the southernmost point. It was Isla Desolación, an island well named. We came back when we did because by that time we had nothing in particular left to eat. When we got back the people turned pale and crossed themselves. They all believed we had been dead for a long time.

I ask the reader's forgiveness and patience for this tedious personal parenthesis.

The Roaring Forties are well known and often luridly described by timid merchantmen and passengers. They carry ships round the Cape of Good Hope to Australia and New Zealand. But Cape Horn passage is a long way south of that – down in the Fearsome Fifties – within danger at times of the Ice-blink, the Floe, and the Bergs. A region of Whales, Sea-leopards, Albatrosses, Penguins, and Phalaropes. So the fearful sea comes rolling round, always from the west, with nothing whatever to impede it, until it strikes the Continental shelf just off the southern tip of South America. Outside this the sea is very deep. Inside it is not really shallow enough to bring up much groundswell or confused seaway. Even in Magellan Strait, where I lost my only cap within ten yards of shore, it was interesting to learn from the chart that this expendable article had sunk in 176 fathoms of water. A current of two or three knots flows round Cape Horn from west to east.

In those regions, in a general way, it is safe to say that the weather is almost always bad and the country is completely uninhabitable. To the west of the watershed, on the Chilean side, it blows a full gale most days, a full storm the rest, and a full hurricane for three days in every three months. East of the mountains the immense flat *pampa* stretches out to the Atlantic; beyond it a shallow sea to beyond the Falkland Islands.

But there are days, three or four on end, when the terrible wind mitigates its rage and the huge rollers sink from a furious forty feet to a friendly twenty feet. Sometimes it even blows from east to west, in Magellan Strait, but never for long. In general the summer is stormier than the winter.

When Magellan went through it was all unknown water. He tacked his way through the Strait, with small sea-room, but with no shallows to strand him. On the world map made by Mercator, the Flemish geographer, in 1538 the Magellan Strait is clearly marked, but Tierra del Fuego is made part of the Antarctic Continent.

Drake came through the Strait with a favourable wind. After that he got into the open Pacific and was blown backwards, clean round the Horn and spent some weeks clawing his way out against the fury of the sea, in the little *Golden Hind*. He opined that 'Pacifico' was an unsuitable name for that ocean. He may have been the first captain to round the Horn.

In a general, local, and perhaps paltry way, sailing round Cape Horn offers no more difficulty than sailing round the Isle of Wight in winter. The local sealers, otter-hunters, and others do it unthinkingly in little cutters. You go south from Cape Froward, between the islands, and wait for your weather in perfect shelter. Most of these islands are high enough to give shelter from any wind. Picking your fair spell, you sail round in a few hours, then hold hard round to port under the lee of the land, and so back to Punta Arenas.

Most yachts and small vessels go through Magellan Strait, where there is plenty of good shelter on either side. The water tends to be deep even in small bays, but you can moor between two trees in the smaller shelters. If the wind did not tear those trees away nothing ever could. In those waters inshore navigation is simple and easy. Thick kelp grows from every rock down to forty feet, so you can go wherever it is clear of kelp. This weed is so strong that you can moor a small vessel by it.

Captain Joshua Slocum in the *Spray* went through that way. I met an old German skipper in Punta Arenas who helped him there. 'I vos zay to him, "you take a dog mit you. Zem Indian beobles zey are not zo goot. If you vos strong zey schust leaf you alone, or beg from you. If you vos alone und asleep zey maybe keel you, for vot you haf. Dake a tog!" ' But Slocum didn't like dogs, and wouldn't have one dirtying his deck. But he thought out schemes of rigging dummies to simulate men on lookout.

At the last minute the old German skipper came on board

to hand him a bag with a bushel of carpet-tacks. 'Hier ist your tog!' Slocum caught on quick enough! When he anchored he rigged his dummies – one from the forehatch and another in the cockpit – and scattered the tacks broadcast all over the deck. Early next morning he was awakened by howls, of barefooted Indians treading on these and hastily sitting down on others. Slocum dispassionately kicked them overboard with his heavy leather seaboots. These poor creatures, the most debased, perhaps, of all humankind, were abominably treated by the white men of all nations. They have mostly died out, now. A few Alacalufes, among the Western islands we explored, perhaps still manage to keep life in their bodies by selling sea-otter skins.

When you get perfect shelter from the westerly winds great trees grow up, to about 200 feet in height. The biggest and commonest of these is the Antarctic Beech, a true beech with very small leaves, miscalled *roblé*, locally, which really means oak. But on the unsheltered outer islets, where the furious storms comb them over the whole year round, these *roblé* trees are blown flat, so that no twig is seen an inch higher than any other twig, and the whole tree is forced to crawl along the ground. A rough, rocky little island looks like a dish-cover, or like a barrow in Wiltshire. When you try to walk across it you find yourself walking on branches over a chasm forty feet deep.

In the *esteros* which we were exploring we easily found sheltered anchorages. The climate is so damp that no bush fire ever could take place. So every old tree finally blows or falls down, but never gets burned away. When we went ashore to explore our *Inesplorado* world it was generally as 'virgin' as our advisers had told us, and usually on a 45° ascent.

No forest is impenetrable. This was mostly of a tall tough form of *maqui* heather, or *bruyère* (locally *tepu*) which burned well, and a very pretty small-leaved, long bell-shaped flowered holly (*Desfontainia*). Also little clinging and sometimes thorny shrubs. After fighting one's way through, uphill, for some fifty or sixty yards one got on to an easier pitch. By that time one was absolutely drenched to the skin, with freezing cold water. One learned to carry one's tobacco and matches in an oilskin bag.

When clear of the 'virginity' of the shoreline one reached the

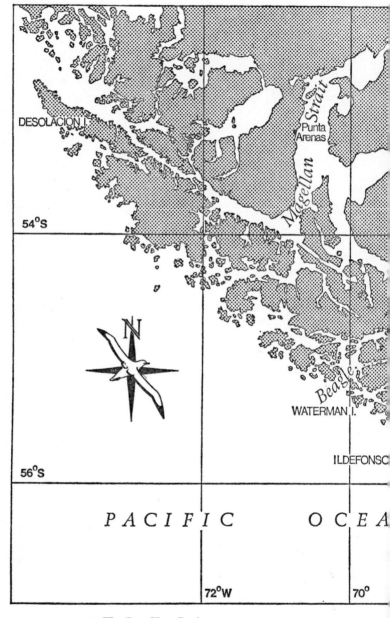

DESOLACIÓN I.

54°S

Magellan Strait

Punta Arenas

N

Beagle

WATERMAN I.

ILDEFONSO

56°S

PACIFIC   OCEA

72°W   70°

7. The Cape Horn Region

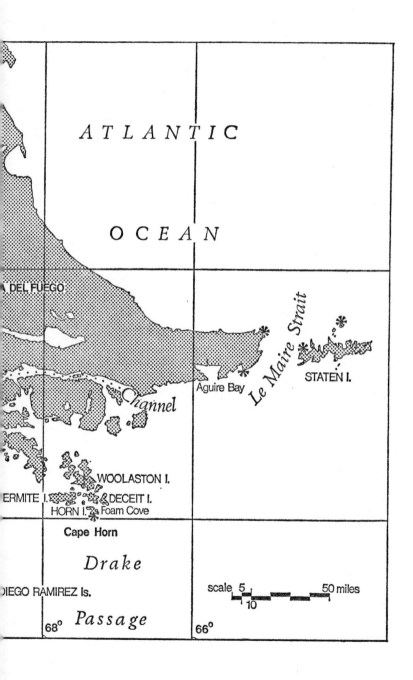

ATLANTIC

OCEAN

A DEL FUEGO

Aguire Bay

Le Maire Strait

STATEN I.

Channel

WOOLASTON I.

ERMITE I. & DECEIT I.

HORN I. Foam Cove

Cape Horn

Drake

DIEGO RAMIREZ Is.

68°

Passage

66°

scale 5   50 miles
        10

collapsible element. Here you find yourself climbing from one fallen tree-trunk to the next for perhaps a quarter-mile. The tree-trunks lie at all angles. The lower ones are presumably the most unsafe, but sometimes an upper one gives way, and you crash through several irregular layers. Then you just have to climb up again. Everything there is dripping wet, and slippery, and moss-encrusted, and unreliable.

So thick and sunless can this forest be that I have climbed from trunk to trunk for five hundred yards without ever clearly seeing the ground, which may have been twenty or even fifty feet below me. On that occasion I bumped my nose into a cliff, having been unable to see it until I actually bumped it. It seemed to go down a long way below me but only about twenty feet up. I went up, and came home by a different way.

Eventually you get above the big timber – quite suddenly, at about 800 feet. Then you are among small cypress trees, which get smaller and thicker until you are treading on what seems like long heather. You are on open ground then, and go rocketing up your mountain like an ibex until you reach the snow. I call to mind one place where a little lake lay between me and the snow. Snow one side, sunshine the other. At 3,000 feet.

The hills above run up to 9,000 feet, with great glaciers pouring down into the sea. With the snowline so low and the discharge from the air so high, it is only natural that the glaciers are huge and overloaded, and that great bodies of ice are continually calved from these. We just had to steer between them. They do not compare with the Greenland or Antarctic Bergs, but would give one a nasty bump. One glacier we explored was eight miles wide, and averaged 200 feet in height, where it met the sea. Some of the icebergs stood thirty feet out of the water.

There were green humming-birds flying about round the foot of that glacier. Bird life in those regions is enchanting. Being a bird-maniac it was delightful to go ashore, upon this beautiful land, and instantly be mobbed by flocks of small birds. Never having seen a human being before they thought we were interesting, or perhaps just ugly. At any rate they crowded round and said so. We loved them. A wonderful land for birds. Not so good for land-mammals. There is a rather inferior deer, called *huemul*, on the mainland and the islands. But they are shy and uncommon. There are a few pumas, a few coypus (nutrias) and a fair number of sea-otters. But the sea-mammals were abundant. The great southern sea-lions were roaring round us every night, however far inland our anchorage. Round the outer islands we would be hounded by sea-leopards who seemed to want to devour our dog. We would see porpoises all the time, and watched penguins fishing, and heard them moo-ing, like little cows. Albatrosses glided round us but never came in between the islands. There are no insects that we ever saw. The climate is too much for them. Even the Indians, whom you can smell a cable-length away, are not lousy. I encountered one small black beetle, several miles inland, who defended himself by giving out an awful stink, which made my hand an offensive liability for hours.

I do not know how many yachts have sailed round the Horn, but when Chichester made his decision to do the passage in a definite length of time he 'put up a job on himself', as Canadians would say. It is easy to duck in to a shelter, within sight of Cape Horn, and wait a few days for a change of weather. But he struck it lucky for weather. In one way a small yacht, especially a yawl or cutter, has it easier than the great square-rigged ships. These, with triple-reefed fore-topsails only, would continually be 'pooped' by the immense following seas. The helmsman – and it sometimes needed two men – had to be protected by a strong iron shelter, as much to stop them looking astern as to prevent them from being washed away. A friend of mine was washed overboard from the foredeck of his ship when she dipped her head into a wave, and the next following sea dumped him, half drowned, on to the poop-deck. He lived for fifty-four years after that.

The sailor of a small yacht, with a small raffee or jib-topsail set high to hold the wind when the vessel is down in the troughs, will keep way on, and be less likely to broach to than a heavier ship with far more top-hamper. Staysails or a try-sail hang slack when in the troughs and then suddenly fill with such a damned slam-bang that even the strongest sails of heavy flax may be torn to shreds, or their halyards part.

The square-rigged ships sailing round the Horn from east to west had to make long tacks, no nearer than seven points from the wind. It was hard going. The captain's main worries were lack of sights from sun or stars, fear of the islands on the northern side, of the ice-blink on the southern side, and of Diego Ramirez in the middle. His crew had troubles too, but these were mainly physical. A small yacht sailor trying to make the same passage would be off his head. At least one long-forgotten ocean racer, to whom many years have taught caution, thinks that way. But a young man of 29 – crazy or not – might try it and might succeed. It is much easier to go through Magellan Strait, where you can find a shelter every night. But, even under comparatively favourable conditions, Magellan Strait isn't just money for old rope. Misadventures can happen there, and sometimes do.

Perhaps my strongest recollection is of some moments of quite serious thought, while sailing round some of the outer islands, in the open ocean. The wind was moderate, but the vast rollers came smashing in upon a shore which was rugged cliffs of rock, hundreds of feet high, with a few broken falls of rocks at their bases. Against those awful rocks the terrible seas smashed their way in, huge, remorseless, regular, unceasing and beautiful. One realized that one was watching a titanic struggle of the irresistible force against the immovable object. Perhaps not quite that, because the sea wins in the end. But it takes a long time. And the glorious savagery of the whole thing left one hoping very seriously that our utterly unseaworthy little boat would not choose that place and time to fall to pieces.

For a young man looking for a very healthy sort of trouble perhaps Cape Horn offers as good as the best. Young men always look for trouble, and generally find it. If they are not manufactured that way they can become drug addicts, or simply shirk life and live on the Unemployment Benefit. These are not really men. They grow their hair long and their fingernails get dirtier not from toil but from continually scratching their own skins.

But for a young man who really is a Man there are many hazardous hobbies, amusements and sports, all of which are worthwhile if danger is not excluded. At the top I would put what that wonderful man Jim Corbett did. He made a life-work of hunting man-eating tigers, in the mountainous and densely forested parts of Northern India. As he pointed out, while you are hunting on steep ground, with thick growth obscuring vision, you know that the tiger you are hunting is at the same time hunting you! Next to top I would put rounding Cape Horn from east to west, without using Magellan Strait, in a small sailing yacht.

The man who hunts the tiger has got to know his jungle, and all the signs of nature, and all about tigers. The worst penalty of failure may be that he is only disabled by the tiger's attack. In that case the tiger may simply eat him alive. This may take two or three days. The Cape Horn young man will know his boat, and his gear, and have proved them against heavy weather. His weak point will be his own capacity to endure what will seem to be eternities without sleep. There comes a point when a young man can fall asleep standing up, and fall down. That would not do off Cape Horn. But his death would be such that he would hardly be conscious of it. Unlike that from the tiger.

The Cape Horn Region has many peculiarities. Once you have known it you cannot ever forget it. The sky is very apt to shine bright green at sunset. No aneroid barometer tells the truth. The stars you knew are gone. You snatch at the acquaintance of Canopus and Fomalhaut but you have no affection for them. The North Star and the Great Bear have disappeared. The Southern Cross you can still see, but it is to northward.

But, when you are very weakened and ill-tempered by hunger and partial exhaustion, you suddenly see about ten thousand flamingoes take the air, from the snow-covered beaches on Tierra del Fuego, and fly over your head on their northern migration. Then you cannot help thinking that this is the most fascinating land and sea that you have ever seen.

20 Small Craft in The Roaring Forties
COMMANDER ERROLL BRUCE, R.N. RTD.

S AILING a yacht round the world is a significant achievement in any circumstances. Yet the Panama and Suez
canals, combined with a programme which allows a
leisurely pace, have brought it within the capability of several
outstanding amateur seamen, especially when supported by a
competent crew.

An open-water circumnavigation, without going through
man-made canals, is a very different matter. It is a much
tougher proposition, even if the course past South America
should be through the Strait of Magellan, then up to the trade-
wind zones and north of Australia; this was done by Joshua
Slocum, who, sailing *Spray*, completed the first small craft
circumnavigation in 1890. Since then Louis Bernicot completed a similar voyage alone in 1938 aboard his 41-foot
Anahita.

Far tougher again is the historic clipper ship route, rounding
the notorious three capes – the Horn, Leewin, and the Cape of
Good Hope. Instead of the moderate easterly trade winds of
the tropical and semi-tropical zones, this passage depends for
its main driving force upon the strong westerly winds of the
Southern Ocean – well called the Roaring Forties.

The first small craft to make this voyage was the 42-foot gaff
cutter *Saoirse*, which Conor O'Brien designed and sailed round
the world, returning to Dublin exactly two years after he had
left in 1925; he started off assisted by a crew of two men,
shipped a new crew at Port Natal before sailing on to Melbourne
and Auckland, then he continued with three men in his crew

for the voyage of 5,800 miles to the Falkland Islands; he made two more calls in the Atlantic before reaching Dublin.

This was recognized by all yachtsmen as an amazing voyage, not only weathering the three notorious capes without damage, but also achieving a speed around the world which was amazingly fast for a yacht of her size.

Some fifteen years later the Argentinian yachtsman Vito Dumas sailed his 42-foot ketch *Legh II* on a voyage which would have astonished all the world had not most of it been struggling in war. Sailing alone from Buenos Aires in June 1942, he called at the Cape of Good Hope, then made a record non-stop voyage of 7,200 miles in 104 days past Australia and on to Auckland; next he sailed to Valparaiso, and made a final voyage of 3,000 miles in 71 days, doubling Cape Horn to reach Buenos Aires 13 months and 11 days after he had set off.

He was the first man to sail alone on a circumnavigation south of the three capes; he had achieved easily the fastest single-handed voyage round the world, and also by far the longest non-stop solo voyage between ports of call.

No long distance voyage could compare with this feat by Vito Dumas for another quarter of a century. Then in 1966 two separate yachtsmen set out alone from Britain, each planning to sail even faster round the world on the clipper ship route.

The first to leave was Alec Rose in *Lively Lady*, but a collision in the English Channel sent him back to harbour. Thus that year it was Francis Chichester alone, and he set out in his *Gipsy Moth IV*, designed specially for the voyage; he planned to sail single-handed at a speed comparable with the wool clippers of a century before, usually manned by some twenty-five men. Chichester achieved his principal purpose in a manner that staggered everyone with experience of long-distance small craft sailing. His speed round the southern hemisphere, counting the days between re-crossing his track in the South Atlantic, for comparison with Vito Dumas whose voyage had been solely in southern waters, was nearly twice as fast as had ever before been achieved in a small craft whatever the size of her crew; from Britain he took nine months compared with the 24 months for the fine voyage by Conor O'Brien with his series of strong crews. Chichester's longest distance between entering a harbour – 15,500 miles from Sydney to Plymouth – was twice the previous record held by Vito Dumas. Quite apart from Chichester's 65 years and his state of health, this was a staggering feat of pure seamanship and navigation.

To beat these amazing records of Sir Francis Chichester would scarcely be possible in any yacht less able than his *Gipsy Moth IV*, but it is another stupendous feat of seamanship that Sir Alec Rose has sailed his twenty-year-old cruising boat *Lively Lady* around the world on the clipper ship route, far quicker than any other man except Chichester. Rose was also no young man, as he celebrated his sixtieth birthday just after the completion of his voyage.

The longest non-stop leg by Alec Rose was slightly shorter than Chichester's, as damage to *Lively Lady* forced her for repairs into Bluff on the southern point of New Zealand. However, the voyage from England to Australia was about twice as long as Dumas' previous solo record of 7,200 miles. *Lively Lady* was not designed for racing speeds, which meant that Alec Rose was at sea, without any let-up from the constant strain of commanding and manning a vessel, for many more days than any man had previously endured at sea alone.

Alec Rose's voyage is remarkable as a brilliant feat of seamanship, both for its length and particularly the thousands of miles sailed in the rough seas of the Southern Ocean. *Lively Lady* is slow compared with most modern sailing yachts of her size; yet the voyage was made in amazingly fast time by persistence in keeping her driving in terrible weather. The personal achievement of Alec Rose is even more remarkable because he set off in spite of the accidents that would have crippled most plans; he started even when the arrival of *Gipsy Moth IV* had made it quite impossible, bar some miracle, for *Lively Lady* to achieve a faster time.

Rose was under no obligation to sponsors. Indeed, he had none at all when he left England, and his only compulsion was a personal determination to attempt such a difficult voyage. The same *Lively Lady* had taken him single-handed across the Atlantic three years before, a feat which inspired the feeling that he and his boat could take on a far more demanding task.

The climax of this voyage was the rounding of Cape Horn, and perhaps the most difficult aspect of this was that it came after sailing many thousands of miles in the rough seas of the Southern Ocean. Rain, snow and fog often made navigation deplorably difficult; an accumulation of defects and shortage of sleep made the approach to the Horn an exceptionally severe problem, and Rose well knew that many ships with full crews had been lost in the attempt.

The actual rounding of the Horn is so much easier when a craft sets out from the shelter of the islands of Tierra del Fuego, with her position known exactly and with the opportunity to select favourable conditions. Indeed Bill Watson, when starting in local waters, reported no great problem in exploring Horn Island in a canoe, although he told that the year before he had anchored his *Freedom* in Horn Island's Foam Cove, after a terrifying finish to his 5,000-mile voyage from New Zealand. Driven towards the Horn in bad visibility, Watson told how his nerve was broken by depressive fits and anxiety; so he credits his 40-foot yacht with having herself steered clear of Diego Ramirez Rocks in a south-westerly gale, and then she drove safely past the Horn into the sheltered cove under its lee, where Watson was sick with relief when he found himself safe.

Many more yachts have sailed round the Horn than the tiny handful which have completed the circumnavigation in southern waters. Possibly the first was the 40-foot American cutter *Tocca*, which J. M. Crenston sailed from New Bedford to San Francisco in 1849, taking 226 days for this voyage of 13,000 miles; yet the records do not specify that he doubled the Horn itself, so he could have sailed through the Strait of Magellan.

It is certain that George Blyth of Great Britain, with Peter Arapakis of Australia as his crew, sailed round the Horn in *Pandora*, which was built as a copy of Slocum's *Spray*; she set out from Australia, put into Auckland for repairs, then visited Pitcairn and Easter Islands, before rounding the Horn on January 16th, 1911, in a storm so violent that they saw nothing at all. Their troubles were far from over, as an even worse storm hit *Pandora* near Staten Island, and she turned turtle,

R—M

losing her mast; next day a whaling ship came upon the derelict and towed her to the Falkland Islands for repairs. Yet *Pandora's* ultimate storm came in the North Atlantic, as she sailed from New York for Europe, but never arrived.

Saoirse was the next recorded yacht to sail round the Horn, then came the 36-foot cutter *Mary Jane*, sailed round by Al Hansen, accompanied only by a dog and a cat. He left Rio de la Plata to double the Horn in midwinter 1934, going from east to west against the prevailing westerly winds. *Mary Jane* safely reached Ancud in Chile after a voyage of 100 days, but was lost on the Chilean coast soon afterwards.

Eighteen years later came Vito Dumas, who also rounded the Horn in midwinter as part of his great single-handed voyage round the world, already described. Marcel Bardiaux, in 1952, again selected the southern mid-winter in his 31-foot sloop *Les Quatre Ventres*. Crossing alone among the southern islands, he set out from Aguirre Bay in Teirra del Fuego, and anchored off Deceipt Island, so that he would be able to round the Horn in daylight for photographs; when daylight came he caught only a glimpse of Horn Island through the driving rain as he rounded it on May 12th before running back past False Cape Horn to the Beagle Channel. His worst experience was when beating through Le Maire Strait between Staten Island and Tierra del Fuego, with the Horn still about a hundred miles ahead of him; there *Les Quatre Ventres* turned turtle, but her twenty-four special buoyancy tanks brought her upright, fortunately with her mast still standing.

Another yacht to turn right over in the vicinity of the Horn was the 46-foot ketch *Tzu Hang*, and she did this not once but twice on two successive attempts to sail round the Horn. Sailed by Miles and Beryl Smeeton, with John Guzzwell on board to help them for that part of their voyage round the world, they left Melbourne and sailed south of New Zealand bound for the Horn. She was running before a gale with sixty fathoms of rope trailed astern to steady her, when a great rogue wave threw her stern over bows; *Tzu Hang* came up dismasted, with a gaping hole in the deck where the doghouse had been torn out, and the boat half full of water; Beryl, who had been at the tiller, came to the surface thirty yards away from the wreck. Somehow they kept the yacht afloat, set about temporary repairs and eventually made northward for a Chilean harbour beyond the ravings of the Roaring Forties.

It was the next southern summer before *Tzu Hang* was ready to put out into the Pacific Ocean again. Two weeks later – it was the day after Christmas – as she lay a-hull without sails set in a fierce storm, she was again turned right over. She was rolled over sideways by a great wave, and came up with her mainmast broken off, part of the mizzen mast snapped and the main hatch gone. There was just Miles and Beryl Smeeton on board the waterlogged yacht, lying disabled about seven hundred miles north-west of the Horn.

Somehow they got back again to tell the tale, and Alec Rose knew all about this when he set out, knowing that he might well find himself in just such a situation; but his *Lively Lady* was a good deal smaller than *Tzu Hang*, and he would have to face it all alone.

In the last few years half a dozen small vessels have rounded the Horn besides *Gipsy Moth IV* and *Lively Lady*. Bill Nance sailed his *Cardinal Virtue* from Auckland to Buenos Aires on an amazingly fast passage for a boat only 25 feet in overall length. He met no damage rounding the Horn, but his boat had previously been turned over and severely mauled by the Southern Ocean seas off Australia's Cape Leewin. In 1966 Edward Alcard rounded Horn Island in his 55-year-old *Sea Wanderer*; like Bardiaux his greatest hazard came when making his approach through Le Maire Strait, and also like Bardiaux he spent many months among the channels and islands of Tierra del Ruego after rounding the Horn itself. Major H. W. Tilman sailed his ex-pilot vessel *Mischief*, not round the Horn but past it, on his way from Patagonia to the Antarctic and back.

Also in 1966, Bernard Moitessier and his wife rounded the Horn in their 39-foot *Joshua*, as part of a voyage of 14,216 miles from Moorea to Alicante; this was by far the longest non-stop distance covered by a yacht until Chichester and Rose exceeded this single-handed.

Yet a further rounding of the Horn in 1966 was achieved by Bob Griffiths and his wife in their 52-foot ferro-concrete cutter *Awahnee II*; it was on their second circumnavigation, and after they had rounded Horn Island in a snow squall with a rising gale, that they turned into Beagle Channel, actually passing Edward Alcard's *Sea Wanderer*.

Nine days after Sir Francis Chichester sailed *Gipsy Moth IV* round the Horn alone in 1967, the 31-foot Australian sloop *Carronade* was sailed round by Andy Whall, with two other men on board; she met very bad weather on the approach and 300 miles west of Magellan Strait she was under a bare pole with two heavy warps trailed astern, when a rogue sea flung her over on her beam ends and half filled her. A day or two later, it was March 30th, 1967 – she passed one mile south of Cape Horn, then turned in to explore the Tierra del Fuego islands.

On the last day of 1967 another yacht rounded the Horn; she was much bigger than those already mentioned, but the well known ocean racer *Stormvogel* is still only a vessel of 73-feet overall length, built to a very light displacement. Caes Bruynzeel, a South African born in Holland, had seven men in his crew; between them they represented eight nations, besides a vast deal of sailing experience. The yacht day-cruised through the islands, rounded the Horn from west to east after breakfast, then most of her crew landed for a New Year's Eve sunbathe on Horn Island itself. Yet when she reached Le Maire Strait next day the cruise became a fight against high seas. After rounding Staten Island, Bruynzeel turned back to round the Horn from east to west. There were no picnics that time, as they met a real Cape Horn westerly gale and rounded in thick weather with visibility reduced almost to nil in the driving rain squalls.

Altogether then, seven men including Alec Rose have sailed alone round Cape Horn, and as many more have taken their small yachts round with crews to assist them. This is a select band of seamen indeed, but smaller still is the quartet of single-handed seamen who have circumnavigated south of the three capes – Good Hope, Leewin, and the Horn. The purist could claim the Dumas' voyage was not a true circumnavigation, as although he sailed through 360° of longitude a man on either pole could do this by merely turning right round.

Dumas made a circle of some 20,000 miles round the bottom of the globe, mostly in the dangerous Roaring Forties, but a circumnavigation by sea which passes through two points diametrically opposite entails a voyage about half as far again. This leaves Bardiaux, who completed his voyage in seven years; Sir Francis Chichester, who took nine months, and Alec Rose who returned to Portsmouth eleven months after he began his voyage. All three achieved supreme feats in their different ways.

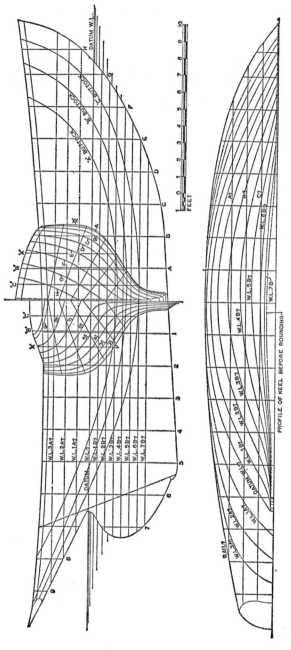

Lines of *Lively Lady*, original design by Fredk Shepherd, M.R.I.N.A.

APPENDICES

Appendix 1
Lively Lady – The Yacht

DOUGLAS PHILLIPS BIRT, A.M.R.I.N.A.

In a technical study of *Lively Lady* it may be of interest initially to compare her in very general terms with the craft of the first three single-handed circumnavigators – Slocum's *Spray* (1895) Pidgeon's *Islander* (1921), Gerbault's *Firecrest* (1923) – and with the immediate predecessor, Chichester's *Gipsy Moth* (1966).

It is a common mistake to believe that the first three of these eminent seamen cast themselves on the ocean in prominently unsuitable craft. Captain Voss may have sailed round the world (not single-handed) in an elderly canoe hacked out of a cedar log by a Red Indian (1896); but his *Tilikum* is an enduring eccentricity among little ships. Of the three ships that first carried lone seamen round the world, two were built for the purpose, while the *Firecrest* was earnestly sought after along the waterfronts of Europe by Gerbault, who was not distressingly short of either time or money. While most circumnavigators have had to be economical, they have set forth on their voyages in craft which were carefully considered and judged suitable by the standards of their day. Of such is *Lively Lady*.

The leading dimensions of the above yachts and *Lively Lady* appear in the table. *Islander*'s and *Firecrest*'s exact dimensions cannot be guaranteed. The dimensions of *Lively Lady* as shown are arrestingly different from those appearing in Lloyd's Register – so much so that someone has suggested that the yacht in Lloyd's must be a different one altogether! They are the measurements taken by the shipwright at Camper and Nicholsons when the cradle for transporting the yacht to London for exhibition was being made. The displacements of *Spray* and *Gipsy Moth* are reliable figures; those of the other three are estimated, fairly rationally it is hoped, and apply to a mean loaded trim.

	Firecrest	Spray	Islander	Gipsy Moth IV	Lively Lady
LOA Ft.	39	36·75	34·0	54	36·0
LWL Ft.	30	32·1	30	38·2	31·0
Beam Ft.	8·5	14·1	10·75	10·5	9·2
Dr. Ft.	7	4·1	5·0	7·75	6·6
Displacement (Tons)	11·0	15·8	11·5	11·5	13·75
Displacement (LWL)³ (100)	410	460	425	200	460

From the technical point of view it is important to appreciate the difference, in terms of testing the yacht (we are not concerned here with the test of the man), between the first three circumnavigations and those of Sir Francis Chichester and Sir Alec Rose. Gerbault's voyage lasted six years. He holds the record for circumnavigation on the instalment plan. Setting out from Cannes, he was the first man to sail the Atlantic single-handed, non-stop, the hard way from east to west. Having reached New York he paused for a year, returning to France while *Firecrest* was 'rejuvenated', in his own word (and she needed it). Thereafter he made his way with dalliances through the South Sea Islands 'In quest of the sun'. He paused again at St Vincent Island and wrote a book. He was back in Le Havre in July 1929, more than six years after his departure.

Setting sail in 1895, Slocum made his circumnavigation in 3 years, 3 months, and 2 days, sailing from Boston (Mass.) and returning to Newport, Rhode Island. Pidgeon sailed on November 18th 1921, from Los Angeles and returned to the same port on October 31st 1925 – say four years. None of these seamen went round Cape Horn; though it might be argued that Slocum, in going through the Straits of Magellan, met dangers enough from the contemporary breed of native around those parts. (The story of the upturned tin-tacks is well known.) Pidgeon and Gerbault used the Panama Canal.

Slocum's *Spray*, as the dimensions suggest, was remarkably un-

usual in form, enormously beamy (a L.W.L. beam ratio of only
2.5), of very heavy displacement, strictly comparable with *Lively
Lady*'s but unlike the latter, with the displacement disposed in a
wide, shallow, dish-like midships section with hard bilges and flat
floors with small deadrise. The contrasting displacement/length
ratios of the five boats appear in the table, the ratio being

$$\frac{Displacement}{(L.W.L.)^3}(100).$$

While of very stiff hull form, *Spray*'s righting moments began
to decrease after 35 degrees of heel, and unlike any of the other
boats considered here she was of capsizable form. Her ballast was
all inside and of concrete. In all respects but the displacement/length
relationship, she makes a singular contrast with *Lively Lady*.

Despite a bow almost as bluff as a Dutch barge and a transom
stern almost as wide as the maximum beam, *Spray* had smooth, fair
lines, and the immersed transom stern gave a sweet run to the
buttock lines of the hull. Above all, the yacht, under a gaff yawl rig
with a large jib set from a long bowsprit, proved to be superbly
balanced. In those days before self-steering gears Slocum was able
to spend days without touching the helm, the reverse of conditions
in the cutter *Firecrest*, which needed constant steering. In this
crucial respect *Spray* was not only the best of the first three single-
handers but, in the absence/of self-steering gear, would have been
superior to *Gipsy Moth IV* and *Lively Lady*, and therefore the best
boat of them all for the lone sailor. She was also excellently built.

Firecrest was the best type of cruiser-racer of her day; but her day
was 1892. Designed by Dixon Kemp, the greatest authority on
yacht design not only at the time but for many years later and con-
tinuing still as an influence, *Firecrest* was emphatically not the type
of boat that most cruising yachtsmen would have selected even
thirty years later as an ocean-going craft for the single-hander.
Gerbault's choice of this very narrow, fast, old cutter, heavily
ballasted on the keel and internally (a ballast/displacement ratio
of possibly 50 per cent), is not the least surprising thing about this
unusual man. Today the sense in the choice is clearer. Initially a
gaff cutter with a very long bowsprit, after reaching New York
Firecrest was, most significantly, converted into a Bermudian cutter,
with the bowsprit shortened, but still long enough. Her rigging and
sails were initially deplorable, thus accounting for the traumas that
enlivened Gerbault's account of his crossing of the Atlantic and make
a most revealing contrast with the Transatlantic voyages of *Lively*

Lady westward and eastward in 1964, which in comparison were quiet sails. *Firecrest* and *Lively Lady*, both unusually well constructed yachts of their day, both heavily constructed and both conventional, the one a cruiser-racer of the 'thirties, the other a cruiser of the 'thirties, differed crucially in the adequacy of their fitting-out, which is more important than anything to do with basic design. The superiority of *Lively Lady* to *Firecrest* lay neither in hull form nor construction, but in much superior rig and rigging.

Pidgeon's *Islander* was in three respects a typical small American yacht: (i) chine hull; (ii) large beam (while L.W.L./beam ratio of *Firecrest* was 3.5 that of *Islander* was 2.8); (iii) she was a build-it-yourself design, and Pidgeon, like Slocum, built it himself. In the U.S.A. the sailing chine hull was accepted long before it became respectable in Europe, and the amateur building of small yachts was popular when it was still a rarity on the other side of the Atlantic. *Islander* was a stock design by Thomas Fleming Day, sponsored by the *Rudder* magazine in which Day ran a well-known monthly column. She was excellently constructed and rigged by Pidgeon, and unlike *Firecrest* set off on her voyage in the best condition. There were no doubt numerous other *Islanders* sailing at the time in Long Island Sound and elsewhere, and while the design was of undoubted merit, the boat does not appear to modern eyes as a desirable one for single-handed circumnavigation. The shallow hull made a long and high coach-roof necessary. She had a counter but a plumb stem, necessitating a bowsprit about 4 ft. 9 in. in length; and there was also a bumpkin. The yacht was rigged as a gaff yawl without topsails, and she carried one headsail on a single forestay from the masthead.

Gipsy Moth was designed by a skilful team of architects – Illingworth and Primrose – with which Sir David Mackworth was associated and which specialized in offshore racing craft, and her design was inspired by the specific purpose of bettering a self-imposed record of speed set for himself by Sir Francis Chichester, derived from the recorded speeds of the last and fastest square rigged ships. The ultimate achievement so far outshone any previous circumnavigation, the voyage extending only between August 1966 and May 1967, with a single port of call at Sydney, that it hardly belongs to the same category.

The basis of *Gipsy Moth IV*'s design was a mainsail not to exceed 287 square feet in area, this being the size determined by Sir Francis as the maximum he could handle, basing the estimate on his previous Transatlantic voyages. The conception of the yacht evolved round this single sail, and to achieve the high average speed required the maximum length of hull possible was necessary. This in turn demanded the lightest practicable displacement, in order to avoid under-canvassing, and a two-masted rig. Thus emerged *Gipsy Moth IV*, much the longest, but also the lightest yacht in the group, and the only yacht of the four designed as a single-handed ocean *racer*.

Of the five yachts of the circumnavigators considered here, four were designed by well-known yacht architects. The origin of *Spray* was more mysterious and as it involved the rebuilding of an earlier vessel we may accept the possibility, suggested by Captain Slocum's son, 'that she had been built in the year one, when Adam was a small boy.' In type, these five yachts box the compass, no greater contrast being imaginable than between *Spray* and *Gipsy Moth*; between *Islander* and *Lively Lady*; but they share the merit of excellent construction, that of *Firecrest* and *Gipsy Moth* by well-known yards to excellent scantlings, of *Spray*, *Islander*, and *Lively Lady* by dedicated amateurs putting more into the work than commerce may inspire. Excepting *Firecrest*, the boats were two-masted, but for different reasons. *Spray* and *Islander* were yawls to improve their self-steering qualities; *Gipsy Moth* was a yawl in order to divide the sail area for easier handling; *Lively Lady* a yawl in order to provide a light weather mizzen staysail.

The Hull

Lively Lady's short ends and relatively narrow beam by the standards of today are characteristic of the once-conventional cruising yacht, and so too is her midship section. She has slightly flaring topsides throughout the length, very slack bilges and full garboards running down to only a little salient keel. It would be difficult to define the separation between the canoe body and the keel. This form of section gives her heavy displacement, which is able to carry her really massive construction. When she returned from her voyage, on the waterline length given above, she was trimming about six inches by the stern. Since Lloyd's differ from the correct dimensions in overall length and beam, one is not encouraged to give much weight to her recorded waterline length there of 27·2 feet; as would be expected, though, she is clearly floating appreciably deeper than designed (as indicated on the drawing), but her generous freeboard has been able to carry the deeper draught without becoming deficient. At her trim on arrival, I should estimate her displacement as about 13·75 tons.

Most of her short overhanging length is in the well-lifted counter, which is curious but not unattractive in shape, without any clearly defined archboard. This may have been a feature that grew into the construction rather than one present in the original design. The sections forward have no considerable amount of flare and there is only slight overhanging length. This has necessitated a short bowsprit, eighteen inches in length. The small bow overhang reminds one of Claud Worth's objection to any great amount of overhang forward; designing today we should certainly prefer to draw the sections out by those eighteen inches, working more flare into the bow and eliminating the bowsprit. That bowsprit, short though it is, might have caused the voyage to end in tragedy instead of triumph.

She was built by her first owner, S. J. P. Cambridge, in Calcutta in 1948. The original design was prepared by Fredk. Shepherd the well-known English architect of cruising yachts, for T. Teasdale, a friend of Cambridge, who as a result of failing eyesight was unable to build her himself, and offered the plans to Cambridge in return for a charitable donation. Then came the War, during which Cambridge made a study of wooden yacht design and construction, and by 1947 he was ready to start building assisted by two Indian cabinetmakers.

Shepherd's design was modified in several respects. Originally her keel, in the manner usual in the 'thirties, had drag aft to the heel of the rudderpost; this was made level to allow the yacht to take the ground more comfortably. A long coach-roof was eliminated, being replaced by a flush deck and skylight, and to compensate for the lost headroom the topsides were raised by about six inches. During Sir Alec Rose's ownership the original main hatch was replaced by the short deckhouse, 4 ft. 6 in. in length. The skylight is immediately ahead of this. The cockpit is deep but small, largely swept by the tiller. As fitted out for the voyage, she has both bow and stern pulpits, the former surrounding the bowsprit, and netting was fitted to the rails from forward to abreast of the mast.

The deck is of laid teak in straight planks joggled into the covering board. The changes made by Cambridge we may regard as wholly to the advantage of the yacht in the light of her future career. Alterations were also made in the construction. The original specification showed steamed timbers, but it was considered that the labour and facilities available could handle grown frames better. One of the most impressive things about the construction is the massiveness of these frames, double throughout, siding 5 in., moulding $2\frac{1}{2}$ in., spaced at $14\frac{1}{2}$-in. centres. As a result the planking is extremely closely supported. Not only this, but the planking,

specified by Shepherd as one-inch thick teak, was actually 1⅜-in.
as laid, the wood ordered from Burma proving on delivery to be
over-thickness. Rather than waste the timber the thickness was
retained. Thus in both planking and framing the original designed
weights were considerably exceeded, producing a yacht of decidedly
heavy displacement, but the additional freeboard enabled the extra
weight to be carried. The result, however, was inevitably a decrease
in the sail area/displacement ratio and a tendency towards sluggish-
ness which was later manifest in practice and led to the sail area
being increased (see below). The seams were splined throughout
and glued with a product of the Dunlop Rubber Company made in
India, and caulking was of cotton and oakum. The timber used for
the frames was Paduak, imported in log from the Andaman islands,
the botanical name being *Pterocarpus dalbergioides* – a timber a little
lighter than teak, and described in *A Concise Encyclopaedia of World
Timbers* – most significantly in the present context – as 'better than
the average as regards movement once seasoned.' It is steadier
than teak.

Perhaps the most remarkable feature of the yacht evident on her
return was the perfect state of her planking. No seams were visible
in the topsides and barely so in the bottom; and the topside enamel,
bright with its turquoise gloss after the long passage, testified to the
steadiness of the construction.

The specified keel was lead, but none being available except at an
exorbitant price, Cambridge redesigned it for cast iron. He
writes of this: 'It was made without a wood pattern. An Indian
moulder was given a box of moulding sand, six full-size cross sections, a small clay model of what the keel should look like, and told
to produce the full size mould. After three days of extreme patience,
using only a small 3 in. trowel, he produced the finest mould I have
ever seen. Little or no fettling was required.'

The finished weight of the keel was 2·67 tons, and there is a further
1·75 tons of internal ballast in lead pigs. On my estimation of the
yacht's present displacement of 13·75 tons, the ballast keel/displace-
ment ratio is thus only about 20 per cent, while the ratio including
the internal lead is 32 per cent.

Below deck the yacht might be described as 'severe', with its
unpainted teak joinerwork of great solidity, and the layout is totally
without gimmicks. It was designed by Cambridge. The galley on
the port side has opposite it a quarter berth, above which is the
chart table. Narrow settees on either side in the saloon have pilot
berths outboard of them; and the foc's'le has two built berths. The

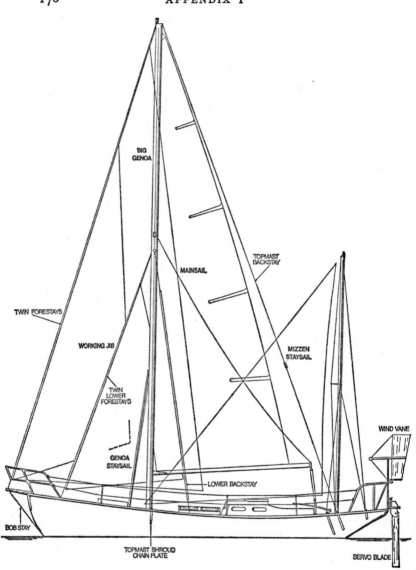

New sail plan of *Lively Lady* designed by Captain John Illingworth.

single-hander has plenty of room in her without that excess of space in which to be thrown far, and compared with the lighter displacement craft of today, with their weight and space saving joinerwork, *Lively Lady* gives the comforting impression below of firm soundness.

The Rig

Lively Lady was completely rerigged by the then firm of Illingworth and Primrose for the Transatlantic race in 1964, and Sir David Mackworth, who was with this firm at the time, was chiefly responsible for the work. He subsequently remained in charge of the rig, and indeed of the yacht generally, for all technical matters. The new rig stood excellently during the Transatlantic race and return voyage, but the yacht was decidedly under-canvassed, and this led to the changes which produced the circumnavigation rig.

Lively Lady was a cutter for the Transatlantic race. In the initial rerigging by Illingworth and Primrose she was given a new mast, by Proctor, stepped on deck, and the basic shroud arrangement then produced was retained in the circumnavigation rig. For the latter the mainmast was lengthened by four feet to increase the sail area and a mizzen mast was stepped. The sole purpose of the latter was to carry a large mizzen staysail, tacked down as far forward as the main mast. This sail was not much less in area than the mainsail, and its object was not to increase the total sail area that might be carried, but to replace the mainsail in light airs, when the former would be slatting and banging rather than driving the ship – those conditions so common on the ocean when there is an old sea but not enough wind, and the fore and aft rig is at its worst.

The lengthened mast of the circumnavigation rig made a total replacement of the standing rigging necessary, and this was all in 7 × 7 construction galvanized wire. (This rope consists of seven strands each composed of seven wires, the seventh strand forming the core of the rope.) Today the usual wire for first-class rigging is of a 1 × 19 construction, the rope being in effect a single strand of 19 wires instead of the 49 individual wires forming the 7 × 7 rope. But 1 × 19 rope cannot be hand-spliced, and though very neat terminal fittings are used, which do not entail the loss of strength caused by a splice, the swages themselves are subject to a deterioration which is hard to detect in its initial stages. The 7 × 7 rope is the strongest that may be hand-spliced, and initially all the ends in *Lively Lady*'s standing rigging were spliced. Subsequently, at Melbourne, the upper ends of the lower shrouds were swaged, and gave no trouble.

The shroud plan of the lengthened mast was the same as that of the Transatlantic mast and is shown in the drawing. It is strikingly simple. A single lower shroud per side is led to a chainplate about 4 feet abaft the mast, this being balanced by the inner forestay. A single spreader and topmast shrouds complete the lateral support of the mast. With the single spreader rig there is a need to steady the long upper panel of the mast, for which purpose a steadying shroud is fitted from the outer ends of the spreader to tangs at the midpoint of the panel. This is not a heavy load-carrying member, its object being only to restrain flexing in the upper length of the mast, not to carry loads due to the sails.

A single preventer (standing) backstay branched in its lower part, the legs taken to chainplates at the quarter, five feet forward of the end of the counter, balance the twin topmast forestays. The lower forestays are also twin. The support of the mainmast thus consists of nine members – four shrouds and five stays. The single lower shrouds, inherited from the shorter-masted Transatlantic rig, were perhaps a concession to the strictest mechanical efficiency, and undesirable in an ocean-going rig. Sir David Mackworth has said, 'Clearly we misjudged the increased pitching loads with the longer mast.' After the rigging troubles experienced on the outward passage, forward lower shrouds were added. Though ideally these should not be necessary, their presence may be regarded as an improvement on the belt-as-well-as-braces principle so desirable in practice in the rigging of ocean-going craft. They are not shown in the plan.

The staying of a tall mizzenless mizzen mast, on which all the sailing loads act forward, is clearly a nice problem, especially in a yacht with a short, tucked-up counter, and compared with the main-mast its staying looks complex indeed. It comprises twin lower shrouds with considerable drift fore and aft; topmast shrouds on spreaders angled forward (one of which was broken on return) and a backstay on each side from the masthead, landed on a chainplate as far aft as possible. It will be evident that neither the straightness nor the security of this mast is of crucial importance, being for use only in light weather and being independent of the rest of the rig. The potential danger of it lies not in its falling overboard but of it damaging the self-steering gear in the process.

One other piece of rigging should be mentioned – the bobstay, which parted on the passage to Melbourne in the process of one of those chain-reactions of rigging failures which are the bane of the modern fore-and-aft rig and the greatest danger that the lone sailor has to expect. The bowsprit is short and heavy with a consider-

able length inboard compared with that outboard. This fact perhaps saved the ship at a crucial moment (see below). The bobstay
parted at the lower large link adjoining the shackle, which made the
connection with the heavy steel stem fitting, a failure possibly
accountable to an imperfect fit of the link and the shackle.

Under the circumnavigation rig the following principal sails were
carried, all of terylene:

Sail	Area (sq. ft.)	Weight (oz.)
Mainsail	285	11
Working jib	182	11
Storm jib	101	11
Genoa staysail	122	$5\frac{1}{2}$
Spitfire staysail	70	10
Light genoa	352	6
Heavy genoa	400	9
Mizzen staysail	230	$5\frac{3}{4}$
Trysail	111	12

The jibs were permanently shackled for the voyage to the twin
topmast forestays, and the genoa staysail and spitfire to the lower
forestays. The light masthead genoas are hanked to the topmast
forestay. These formed the running sails, set with bearing out spars
of light alloy on a system devised by Mackworth. Halyards are run
where possible through large metal eyes on the mast. On her return
the alloy mast had a deep score some five feet long in the locality of
the spreader, where one halyard (nobody is certain which) had sawn
its way into the metal. The result looked more alarming that it was;
for even a longitudinal crack in a mast does not have the weakening
effect that might be expected. Some people will remember the long
rifts that inevitably occurred in the old solid wood masts, which caused
no harm. There is danger of serious weakening, however, if such
longitudinal fault in the mast allows a sliding movement to occur.

Rigging Troubles

It may be of interest to consider briefly *Lively Lady*'s rigging failures
– the account of which appears elsewhere in this book – from the
technical point of view. It should be emphasized that concentration
on the failures must not be allowed to dim the triumph of the rig in
having carried a fine seaman through such a trial as circumnavigation; but it may highlight the inbred weakness of any form of
highly stressed Bermudian rig in the absence of some wonderful
fatigue-free metal.

R—N

The first troubles, as will be known, began occurring in the Southern Ocean, when the yacht was in the most remote position on the oceans of the voyage. It was then, with the genoa about to be hoisted, on the 114th day out, that the starboard lower shroud went at the mast under the spreader. Prior to this failure, as the narrative tells, *Lively Lady* had gone through a succession of gales. The failure occurred at the upper splice round the thimble in the locality of the first tuck; that is, at the top of the splice where the maximum deformation of the strands occurs when making the tucks. If failure were to occur, this is precisely where it might be expected. The immediate remedy, as explained in the narrative, was to lead one of the inner forestays aft and also one of the jib halyards. Later, in a fine effort of seamanship, Alec went up the mast with the repaired end of the shroud, which he had bent round another thimble and secured with bulls-eye grips.

Next the bobstay came adrift from its stem fitting, which it was not possible to repair at sea despite a noble effort. At this point in the great adventure the ultimate security of the ship was thanks to the bowsprit being so short and stout and able to continue some sort of duty when shored from above by a strut. But, of course, a longer overhang and no bowsprit would have removed the possibility of this possibly fatal danger. Bowsprits and their attendant bobstays are not suitable parts for modern Bermudian sail plans with masthead rigs. They simply add one more vital but avoidable potential weakness to the many others.

Then the port lower shroud went the way of the starboard shroud a fortnight later. There was thus a chain reaction of failures; first the starboard lower shroud, on November 7th, followed ten days afterwards by the bobstay on the 17th, and then the port lower shroud four days later on the 21st. The sequence of events – for what comfort this may be – was at least logical, the trouble originating without doubt in violent pitching for days on end, as severely as at any time during the spells of calm. These are the very conditions to produce the sudden and persistent reversals of loading that cause fatigue failures in the weakest links; initially, and not surprisingly in this case, a splice.

At one stage during these troubles, Alec let go from forward one of the topmast forestays and led it aft, round the spreader, setting it up again with a bottlescrew on to the chainplate. This necessary emergency measure quite evidently would have put an unfair bending load on the masthead tang, which may have contributed to its subsequent failure on the return voyage. But even without this, the tangs of fore and aft rigging are the most dangerous element in the hardware. Whereas the tangs of the shrouds may be kept reasonably free from lateral movements and bending moments, fore-and-aft rigging, and most particularly masthead stays carrying sails, tend to set up bending moments in the tang which cause it to work harden and fatigue.

The capital weakness of the modern aerodynamically efficient Bermudian rig – as developed for ocean-going purposes primarily in the offshore racing craft, it should be remembered, which operate under such widely different conditions from those facing the lone sailor – is its highly stressed character and ultimate dependence upon many small metal parts, any one of which giving up the ghost may imperil the entire rig. For this reason alone the modern Bermudian rig is horizon-down in the distance from any ideal for the single-handed ocean navigator. Its secondary weakness lies in the fact that the rig has been developed primarily for improved weatherliness, the quality which brings racing success but is of relatively little value to the circumnavigator. On the other hand, the ability to make efficient use of free winds, the prime need of the ocean sailor, is the one respect in which the Bermudian rig is notably deficient. Yet despite these facts, the contemporary absorption in single-handed ocean sailing has produced no clear rival to a modified Bermudian rig, though the junk rig as developed by H. G. Hasler may prove to be the yet missing device.

The moment of truth will come to the single-handed seaman in open water not probably in the shape of the mountainous waves enjoyed by the popular imagination, but in the creeping, quiet weakening through fatigue of a little bit of rigging that nobody but a specialist might notice, and even a specialist fail to detect in time. Alec's first moment of truth, technically, came when the bobstay went and the whole rig might have gone overboard but for the bow-sprit being short and stout and behaving, under a bending moment, in a way that no engineer would plan for it to behave; for a bowsprit, if you must have one, should be a strut under compression. The second was when the tang on the masthead fitting for the topmast forestay broke; and both troubles may be attributed to fatigue in little metal parts.

Appendix 2

Lively Lady's Wind-Vane Steering

H. G. (Blondie) Hasler, the designer of the gear

Lively Lady's vane gear is a standard production model known as the 'Type 1 FQH', and is made and sold by M. S. Gibb Ltd., Warsash, Southampton, England. It operates on the 'pendulum servo' principle, which can be briefly described as follows.

The *servo blade* S (Fig. 1.) resembles a dinghy's rudder, and hangs vertically over the stern. It is carried on bearings inside the *servo box* F, and can be turned like a rudder by means of the *servo tiller* A. The *servo box* F is itself carried on *fore-and-aft bearings* E which are mounted on the portable *tubular bumkin* B, so that the box F can swing from side to side like a pendulum, taking the blade S with it.

A *servo quadrant* P is integral with the box F, and *steering ropes* W (Fig. 2) lead from it through *steering sheaves* C to the *rudderhead quadrant* Q, which is secured to the top of the tiller.

The plywood *wind-vane* V is connected to its linkage through the *latch gear* L which enables the desired course to be selected, and also enables the vane to be unlatched in an emergency by pulling on a latch line which is led to the cockpit, thus permitting instant reversion to manual steering while the vane 'weathercocks'. When the latch L is connected, any turning movement of the vane V is transmitted via the reversing linkage and the *servo tiller* A to the *servo blade* S. *Please note* that Figs. 1 and 2 are purely diagrammatic, and do not attempt to show the true proportions or layout of the components.

To set a course, the vane is first unclutched at L and the yacht steered manually on to the correct course while the vane weathercocks. The vane is then clutched in at L, and the tiller allowed to swing freely. Fig. 2 shows the vane set for running before the wind, and it will be seen that if the yacht now yaws to starboard, the

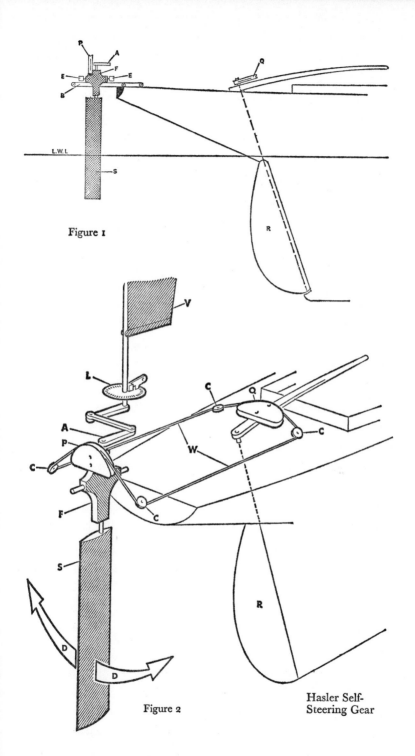

Figure 1

Figure 2

Hasler Self-
Steering Gear

change in direction of the apparent wind will cause the vane V to turn anti-clockwise, when looking down on it. This movement causes the servo blade S to turn clockwise, whereupon the flow of water past it (caused by the forward motion of the yacht through the water) makes it swing sideways to starboard as shown by the right-hand arrow D, thus pulling on the starboard steering rope which turns the main rudder so as to bring the yacht back on course.

It will be seen that the system is a true servo, in that the vane is used to monitor the direction of the apparent wind, but the power used to turn the yacht's main rudder is derived from the water flow rather than from the wind. To turn the main rudder directly by wind power would call for a wind vane at least five times as big.

In practice, a good vane gear of this sort will steer the boat adequately on all points of sailing, and in all wind strengths provided that she is making at least a knot through the water. It is never necessary for the crew to steer a course manually, but – except when beating to windward – it is necessary for him to reset the latch gear every time the direction of the true wind changes, since the yacht will otherwise follow the wind instead of following a desired compass course.

Running or broad-reaching in fresh winds and a big following sea it is not usually possible to drive the boat really hard without experiencing excessive yawing, and it is then necessary to shorten sail a bit in order to slow her down to a reasonable 'cruising speed'. The point here is that a good human helmsman can *anticipate* a yaw, whereas the vane gear cannot.

One useful feature of this type of gear is that even in a full gale it can be completely unshipped, and shipped again, by one man working in safety inside his guardrails; this makes it fairly simple to fit a replacement part at sea. The weight of the complete gear is about 65 lbs.

This 'pendulum-servo' type of vane gear is not by any means the only system in use. Perhaps the most popular of the other systems is the 'trim-tab' type, in which the wind vane turns a small rudder (the Trim Tab) which is itself hung on the trailing edge of the main rudder. The main rudder is left free, and when the trim tab is turned in one direction the force of the water flowing round it causes the main rudder to swing in the opposite direction, and so steer the ship. This is, of course, another true servo system, in which the power of the water-flow is harnessed to turn the main rudder. Trim-tab systems are most easily mounted on external rudders, rather than on internal rudders whose stocks pass upwards through a trunk in the hull.

Of the other yachts which are known to have sailed round the
Horn with our vane gears, Sir Francis Chichester in *Gipsy Moth IV*
carried a large pendulum-servo gear, virtually identical with that
used by Geoffrey Williams in *Sir Thomas Lipton* to win the 1968
Single-handed Transatlantic Race, whereas Bill Nance in *Cardinal
Vertue* used a trim-tab gear on his transom hung rudder.

All vane steering gears are of fairly recent development, having
first been demonstrated as a practical device in the model yacht
racing classes in the early 'thirties. In 1936, the French transatlantic
yachtsman Marin Marie devised and made up a workable full-
sized vane gear while crossing the Atlantic from west to east in the
45 ft. motor cruiser *Arielle*, but it was not until 1955 that effective
vane gears first appeared on full-sized sailing boats: Ian Major's
Buttercup, and Michael Henderson's *Mick the Miller*.

The value of vane gear on a long voyage is obvious, but most
owners of vane gears have no intention of crossing oceans; they
use their gears in ordinary coastal cruising to save having to sign on
unwanted crew members. Looked at this way, the gear is the
equivalent of at least one extra man, but with the advantage that
most vane gears do not eat, answer back, or woo other people's
girls in harbour.

Lively Lady's vane gear may perhaps have steered her for a greater
distance than any other vane gear yet built. When Jock McLeod
and I first went over to Attrill's yard at Bembridge in December
1963 to advise on the fitting of the gear, all we knew was that an
unknown Mr Rose had bought the boat with the intention of
entering for the 1964 Single-handed Transatlantic Race for the
Observer Trophy, and had commissioned Illingworth and Primrose
to re-rig her for single-handed sailing.

Three months later we sailed trials in the Solent with Alec, and
found him so quiet and retiring that we could not really decide
whether he fully understood what he was tackling, or not. These
doubts began to be dispelled a few days after the start of the race,
when by extraordinary coincidence my own boat *Jester* and *Lively
Lady* found themselves converging close to each other out of an
empty sea on three separate occasions, and I could see that Alec
always seemed to have her beautifully under control, tramping
along with the right number of reefs down and an obvious absence of
drama. When I finally got to Newport, Rhode Island, it was no
surprise to find that he had finished 4th out of 15 starters, and over
a day ahead of me.

He then made a very good single-handed passage back to England,

and it was with this same veteran steering gear that he started off round the world in 1967, leaving us anxiously wondering what components were going to wear out first, and whether he had taken enough spare parts.

Our biggest relief came when he was photographed off the Horn with the gear still working, and when it finally brought him up the English Channel we were able to see for ourselves what signs of age it was showing after steering for a total of at least 36,000 nautical miles of open water, and how well Alec had looked after it.

As designers, we have learned a good deal from his marathon performance, and would like to offer our congratulations, and thanks.

Curdridge   H.G.H.
26th July 1968

Appendix 3

Stores and Provisioning

ALEC ROSE

Provisioning ship for a voyage such as this was an important operation. My previous experience in sailing single-handed around the coast, across the English Channel, and the Transatlantic Race stood me in good stead. The problem was that on this occasion, a far greater reserve stock of food and water was needed. The voyage could last up to twice as long as anticipated. One could get into trouble thousands of miles from port and right off the shipping lanes, involving a struggle of perhaps months in making port.

I have got into a routine with meals, which I have always followed as far as is possible when sailing single-handed. I also try to eat sensibly. That is, food that is varied and sustaining. A typical day's diet would be to start off at dawn with a hot cup of tea with perhaps a dash of whisky. Breakfast would consist of a cereal such as Shredded Wheat or porridge with a handful of raisins in it, with brown sugar and hot milk, followed by fried eggs, with fried potatoes and onions with bread or crispbread biscuits, finishing off with an apple or orange. A hot coffee or cocoa during the forenoon would be followed by a good lunch of perhaps a tin of stewed steak to which would be added baked beans, tomatoes, carrots, onions, plus a pinch of dried mixed herbs for flavouring. This would last two days. Potatoes I always cooked separately and enough to have some left over to eat cold with cold meat or salmon or something of the kind, washed down with a can of beer. I would also fry them up for supper with baked beans, tomatoes, and eggs, or sausages. A cup of tea, with a slice of fruit cake in the afternoon, would be followed by supper of cheese, butter, and biscuits with a raw onion, buttered biscuits with honey and a slice of fruit cake. After every meal I had fresh fruit, such as an apple, orange or, banana while they lasted. Last thing at night I would have a cup of Ovaltine or cocoa with plenty of milk.

During the night or at any time I felt the need of it I would have my favourite hot drink of a squeezed lemon, honey with hot water and topped up with whisky.

I tried to plan my diet to keep me healthy and regular in my habits. After storing up with such items as tinned goods and packet cereals, the next thing on my list of priorities, would be fresh fruit and vegetables. Potatoes were stored in small sacks in the foc's'le, also carrots and onions. I found they kept very well, onions especially. They had grown long green shoots before I had finished them but they were delicious with biscuits and cheese – just like green onions. They lasted to within a week of reaching Melbourne. Similarly, onions put on board in Australia lasted all the way back to England. Greenstuff in the way of lettuce and cabbage last up to a fortnight and tomatoes, if obtained slightly unripe, will keep three of four weeks if picked over and eaten as they ripen. Tomatoes put on board at Bluff, New Zealand, lasted over five weeks without one going bad. Of course, the weather was cold down in those latitudes. I had two cases of Australian apples, and one case each of lemons, oranges, and grapefruit. I left them in their original boxes, and stored them in the foc's'le. Hardly any were thrown away and I actually had one lemon left when arriving in Australia.

I took brown wholemeal bread with me, which I put into sacks so that the air could circulate round it. It kept for about three weeks, going green on the outside, but, after cutting this off, it was quite good inside. Ten dozen fresh farm eggs were taken. They were not treated in any way. Previous experience had proved that they will keep as long as is necessary without it. Sterilized milk in sealed bottles will keep for months and I used this for making Ovaltine or on cereals. Tinned milk I used for porridge or in coffee, and Marvel powdered milk in my tea. Small whole cheeses are good to take. They keep well if exposed to the air and are good food value.

I cooked by Calor gas. Many people, I know, would not have gas on board, but I found it very good and it is safe if used properly. I took two spare cylinders with me which proved adequate. I had a paraffin lamp for lighting as well as a Tilley storm lantern and a Tilley heater. Of course, I had electric lamps as well but these were used sparingly or for chart work only in order to save the batteries. About ten gallons of paraffin was taken and this I stored in bottles in the cockpit lockers and in the bilge. It is convenient to fill a lamp out of a bottle when rolling about at sea.

About a hundred gallons of water was carried altogether. In addition to about sixty gallons in the tanks, forty gallons was put aboard in plastic jerry cans. I made no provision for catching any rainwater, as I felt I had adequate supplies.

I give below a list of the main stores put on board before leaving England, to last the whole passage round the world, except for replenishments at Melbourne, my one proposed port of call. They were:

Brown sugar	35 lb.
Tinned Chicken Suprême	84 tins
Salmon	41 tins
Creamed rice	21 tins
Honey	12 jars
Golden Syrup	5 tins
Stewed steak	22 tins
Porridge oats	20 packets
Shredded Wheat	24 packets
Cocoa	3 tins
Ovaltine	6 tins
Complan	8 tins
Cakes	24 × 1 lb. tins
Tea bags	3 gross
Dried eggs	84 small tins
Sausages	32 tins
Milk	44 tins
Corned beef	45 tins
Butter	12 × 1lb. tins
Baked beans	18 × 1 lb tins
Baked beans with sausages	12 × ½lb. tins
Sardines	42 tins
Herrings	10 tins
Steak puddings	18 tins
Soups	20 tins
Cheeses	6 × 1 lb.
Bovril	1 jar
Biscuits (cheese)	3 doz. packets
Biscuits (sweet)	100 small packets
Biscuits (shortbread)	3 × 1 lb. tins
Potatoes	12 × 1 lb. tins
Carrots	24 tins
Nescafé	5 tins
Pre-packed curry	48 packets
Pre-packed rice	48 packets
Ryvita crispbread	6 packets
Energen crispbread	12 packets
Instant mashed potatoes	24 packets
Heinz beef casserole	24 tins

Heinz self-heating soups	24 tins
Sultanas	3 packets
Beer	100 tins
Gin	6 bottles
Whisky	8 bottles
Brown loaves	24 × 1 lb.
Eggs (fresh)	10 doz.

Also, apples, oranges, grapefruit, lemons, bananas
tomatoes, lettuces, onions, potatoes and carrots.

On leaving Australia for the return voyage, all the fresh fruit and
vegetables were replenished free of charge to me, as were the fresh
eggs, fresh bread, and much tinned food.

Similarly on leaving Bluff in New Zealand, fresh fruit was given
me as well as bread and many other gifts such as jam, biscuits,
whisky, and canned beer.

I arrived back in England with a good stock of food still on board
and I think I carried too much weight around with me – but, as I
have said, one has to provide for all eventualities.

Appendix 4

Weathering Storms

ALEC ROSE

'What do you do in a gale?' Many times I have been asked that question. It is difficult to give a straight answer – in fact, almost impossible, as conditions at the time vary so much. Sea conditions, such as wave formation, height and regularity of waves, count as much as the strength of the wind – in fact more so, I think, as it determines what course of action to take such as:

(*a*) Running with it at speed
(*b*) Heaving-to
(*c*) Towing warps
(*d*) Lying to a sea anchor
(*e*) Lying a-hull

As regards (*a*) I always try to run as long as I can, providing it is in the right direction, of course. I have done this on many occasions with the wind dead aft and with a boomed-out headsail. It is exhilarating sailing, with the yacht surfing at times as a following wave breaks under her and carries her along with it. But there comes a time when the yacht begins to get out of hand as the seas build up with the wind. Sailing single-handed there is a limit to one's endurance in staying in the cockpit to steer. I found that my self-steering gear was at times unable to cope, if a cross wave swung her round off course, and I had to be on hand to put her back. One's hearing plays an important part on those occasions, to judge the strength of the wind by the noise in the rigging. I found nearly always that it was the sea that forced one to reduce sail while running, not the strength of the wind. It was those awkward cross seas that come aboard on the beam, with the roll *Lively Lady* develops while running at speed; they fill the cockpit up and, hitting hard at times, swing her stern round. It was then that I would reduce sail to a small staysail and keep going on that, or a very small storm jib. I could not get *Lively Lady* to run under bare poles at all. I tried often, but she would always end up lying a-hull with no way on her at all. Of course, with a strong crew and sound gear a modern yacht can be kept going much longer than a single-hander does.

This brings me to (*b*), heaving-to. When the gale develops from ahead, and it is no longer a proposition to beat against it, I have usually hove-to. By that time I would have been under shortened sail anyway, so it was a simple job. I found that she would heave-to readily under reefed mainsail and a headsail or headsail or mainsail alone. During a southerly gale off the Tasmanian coast I headed offshore for the night under reefed mainsail alone. She lay quite comfortably, making slight headway. At other times I hove-to under a staysail. Heaving-to has its advantages over lying a-hull in that the yacht is more steady. In the violent storms I experienced in the Southern Ocean, however, there comes a time when it appears that the sails are in danger of being carried away, and gear damaged, or ultimately that the yacht would be laid flat. More than once I had to crawl along a sloping deck with seas breaking over to lower a staysail, under which I was hove to, the yacht being laid well down by the pressure of the wind. The danger of losing sails was always on my mind in a voyage such as this. The problem of chafe was always there and the strain on the clew of a sail hauled to weather is great. To heave-to in order to avoid a hard beat to windward in a strong gale is perhaps wise, but when it gets to storm force I always feel it is better to take all sail off.

Running off before the wind (*c*) and towing long warps astern is advocated by many. I kept a long rope permanently coiled and lashed to the guard-rail aft, ready for use in this way, with another heavy one coiled in the cockpit as well, but I never got to streaming them. Some might say that they would have made the yacht more steady while running under headsails, but with the size of the seas I doubt it. The big seas would pick *Lively Lady* up bodily and swing her round. I am sure the ropes would have gone with her. I do not think they would stop the yacht from being turned over, either. The great waves of the Southern Ocean would lift the rope over as well. Miles Smeeton found this when he was tipped stern over in his yacht, *Tzu Hang*, whilst attempting to round Cape Horn. He was towing sixty fathoms of heavy three-inch rope.

Lying to a sea anchor (*d*) has largely gone out of fashion. I carried a heavy one with me, but never found the need to use it on this voyage. I did use one once, when caught out in a sharp gale in the Bay of Biscay while cruising with my wife in my ketch *Neptunes*

Daughter. I streamed it from the stern first and she lay with the weather just aft of the beam. Then I moved it forward and she lay with the weather just forward of the beam. But it would not hold her head up to the wind. Mind you, the strain taken by the sea anchor on the rope was terrific so it must have stopped a considerable amount of drift. In close proximity to a lee shore, a sea anchor might be a saving factor, but in the open sea I am not convinced of its usefulness. In some of the storms I experienced I think, on the contrary if it had not been carried away it might have damaged the yacht by dragging her through the crest of the wave. Captain Voss was the great advocate of sea anchors, and the Voss sea anchor was a must for most cruising men at one time. But he was overwhelmed and dismasted while lying to one in the Pacific.

Which brings me to my theory that it is safer to lay a-hull (*e*). That is to strip off all sails and let the yacht go with the sea and take up her own position. In this way *Lively Lady* would lay with the weather forward of the beam and with a list to leeward. Big seas coming along and breaking over her, she would give to them and go with them. She would fore-reach slightly. Of course, some would hit her very hard and we were thrown about unmercifully – but no more than we would have been in other circumstances. The fact that the yacht could give to the seas was the strong point, I think. A strong, well built boat is required, of course, but then to venture down into those waters, or indeed any other ocean passage, in anything else is foolhardy. Many cases are on record of yachts having weathered gales safely after the crew have become too exhausted to work the boat, and have left her to herself, or have even abandoned ship. My usual practice is to have a hot drink and turn into my bunk, but I must admit that this was not always so. On some occasions it was so bad I stood by the hatchway in my oilskins and wondered how much more she would take. Boiling foam, seething across the deck, fills the cockpit and the fierce wind lays her over, even under bare poles. In these cases, though, it is difficult to see what else could be done. To get even a tiny sail up was impossible and even if it were, to run in those conditions would, I think, have been dangerous, as the seas hit even harder when the yacht is moving.

On the whole, then, I think that in the very fiercest storms I shall continue to lay a-hull. Unless of course I am caught on a lee shore, which God forbid. I know many will disagree with me but as I said, it is difficult to be dogmatic about it. It all depends on conditions at the time and the type of boat one has. One thing is sure, though. Whatever you do, you will have plenty of armchair critics.